Florian Plentinger

Systematic Model Building with Flavor Symmetries

Florian Plentinger

Systematic Model Building with Flavor Symmetries

A New Approach

Südwestdeutscher Verlag für
Hochschulschriften

Imprint
Any brand names and product names mentioned in this book are subject to trademark, brand or patent protection and are trademarks or registered trademarks of their respective holders. The use of brand names, product names, common names, trade names, product descriptions etc. even without a particular marking in this work is in no way to be construed to mean that such names may be regarded as unrestricted in respect of trademark and brand protection legislation and could thus be used by anyone.

Publisher:
Südwestdeutscher Verlag für Hochschulschriften
is a trademark of
Dodo Books Indian Ocean Ltd., member of the OmniScriptum S.R.L Publishing group
str. A.Russo 15, of. 61, Chisinau-2068, Republic of Moldova Europe
Printed at: see last page
ISBN: 978-3-8381-2415-5

Zugl. / Approved by: Würzburg, Julius-Maximilians-Universität, Diss., 2009

Copyright © Florian Plentinger
Copyright © 2011 Dodo Books Indian Ocean Ltd., member of the OmniScriptum S.R.L Publishing group

Preface

The observation of neutrino masses and lepton mixing has highlighted the incompleteness of the Standard Model of particle physics. In conjunction with this discovery, new questions arise: why are the neutrino masses so small, which form has their mass hierarchy, why is the mixing in the quark and lepton sectors so different or what is the structure of the Higgs sector. In order to address these issues and to predict future experimental results, different approaches are considered. One particularly interesting possibility, are Grand Unified Theories such as $SU(5)$ or $SO(10)$. GUTs are vertical symmetries since they unify the SM particles into multiplets and usually predict new particles which can naturally explain the smallness of the neutrino masses via the seesaw mechanism. On the other hand, also horizontal symmetries, *i.e.*, flavor symmetries, acting on the generation space of the SM particles, are promising. They can serve as an explanation for the quark and lepton mass hierarchies as well as for the different mixings in the quark and lepton sectors. In addition, flavor symmetries are significantly involved in the Higgs sector and predict certain forms of mass matrices. This high predictivity makes GUTs and flavor symmetries interesting for both, theorists and experimentalists. These extensions of the SM can be also combined with theories such as supersymmetry or extra dimensions. In addition, they usually have implications on the observed matter-antimatter asymmetry of the universe or can provide a dark matter candidate. In general, they also predict the lepton flavor violating rare decays $\mu \to e\gamma$, $\tau \to \mu\gamma$, and $\tau \to e\gamma$ which are strongly bounded by experiments but might be observed in the future.

In this book, we combine all of these approaches, *i.e.*, GUTs, the seesaw mechanism and flavor symmetries. Moreover, our request is to develop and perform a systematic model building approach with flavor symmetries and to search for phenomenological implications. This provides a new perspective in model building since it allows us to screen models by its predictions on the theoretical and phenomenological side, *i.e.*, we can apply further model constraints to single out a desired model. The results of our approach are, *e.g.*, diverse lepton flavor and GUT models, a systematic scan of lepton flavor violation, new mass matrices, a new understanding of lepton mixing angles, a general extension of the idea of quark-lepton complementarity $\theta_{12} \approx \pi/4 - \epsilon/\sqrt{2}$ and for the first time the QLC relation in an $SU(5)$ GUT.

Contents

Preface 1

1 Introduction 3

2 Status Quo – Experiments and Theories 7
   2.1 Phenomenological Status 7
   2.2 Theoretical Approaches 10
   2.3 Summary 12

3 Flavor Symmetries 13
   3.1 A Primer to Flavor Symmetries 14
   3.2 Discrete Flavor Symmetries 15
   3.3 Summary 18

4 Flavored Models 19
   4.1 Froggatt-Nielsen Mechanism 19
   4.2 A $Z_5 \times Z_9$ Lepton Flavor Model 20
      4.2.1 Notation and Model Outline 20
      4.2.2 Lepton Masses and Mixings 23
      4.2.3 New Sum Rules 25
   4.3 Summary 26

5 Textures – A Bottom-Up Approach 27
   5.1 Textures with Extended Quark-Lepton Complementarity 28
      5.1.1 Extended Quark-Lepton Complementarity 28
      5.1.2 Mass Matrix Production 28
   5.2 Essence of EQLC-Mass Matrices 32
      5.2.1 Seesaw Realizations 32

	5.2.2 Performance	42
5.3	Summary	44

6 Lepton Flavor Models — 47

- 6.1 Flavor Structure . . . 47
- 6.2 Textures Becoming Models – A Group Space Scan . . . 48
- 6.3 Summary . . . 54

7 Lepton Flavor Violation — 55

- 7.1 Charged LFV in SUSY . . . 55
- 7.2 LFV for CP Conserving and CP Violating Textures . . . 57
- 7.3 Results for LFV-Rates . . . 59
- 7.4 Summary . . . 61

8 5D SUSY $SU(5)$ GUTS with Non-Abelian Flavor Symmetries — 63

- 8.1 4D SUSY $SU(5)$ GUTs with Abelian Flavor Symmetries . . . 63
- 8.2 Scanning SUSY $SU(5)$ GUTs with Abelian Flavor Symmetries . . . 66
- 8.3 5D SUSY $SU(5)$ GUTs with Non-Abelian Flavor Symmetries . . . 67
- 8.4 Summary . . . 70

9 Summary and Outlook — 73

A Appendix — 77

- A.1 Supplementary Information for Seesaw Realizations . . . 77
- A.2 Supplementary Information for Lepton Flavor Models . . . 80

List of Figures — 83

List of Tables — 85

Bibliography — 87

Chapter 1

Introduction

In the area of high energy physics, symmetries demonstrate to be the fundamental concept to explain the behavior and properties of particles. This is consolidated in the Standard Model of particle physics (SM) with its gauge group $SU(3)_C \times SU(2)_L \times U(1)_Y$. However, some open questions remain such as: Why do we have three generations of particles with such a strong mass hierarchy in the quark sector, why is the mixing in the quark and lepton sectors so different, or what is the structure of the Higgs sector?

In order to address these open questions, different approaches are considered in literature as illustrated in Fig. 1.1. They lead to extensions of the SM such as supersymmetry (SUSY), extra dimensions and Grand Unified Theories (GUTs). Nevertheless, the SM describes very well

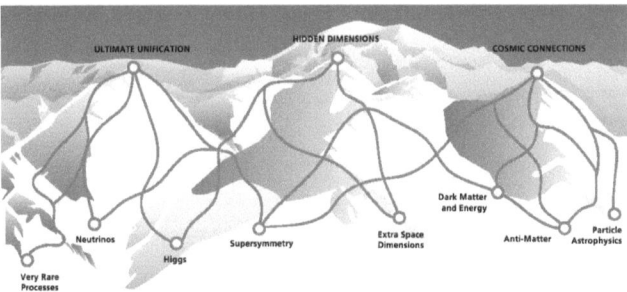

Figure 1.1: Different roads to Grand Unification (taken from Ref. [1]).

the world of known elementary particles. Only recently, solar [2], atmospheric [3], reactor [4], and accelerator [5] neutrino oscillation experiments, have very well established that neutrinos are massive. This astonishing result was the first clear evidence[1] for physics beyond the SM, where neutrinos are predicted to be massless. Moreover, the smallness of neutrino masses $\sim 10^{-2} \ldots 10^{-1}$ eV can be naturally explained by GUTs [6–8] via the seesaw mechanism [9–17]. Thereby, the absolute neutrino mass scale becomes suppressed by the B-L breaking scale $M_{\text{B-L}} \approx 10^{14}$ GeV which is close to the GUT scale $M_{\text{GUT}} \approx 2 \times 10^{16}$ GeV [18–20]. This, on the other hand, means that the seesaw mechanism connects high energy physics with low energy experiments.

However, these are not the only features making neutrinos unique and bringing them to the focus of actual research. Current neutrino oscillation data shows that the Pontecorvo-Maki-Nakagawa-Sakata (PMNS) lepton mixing matrix U_{PMNS} [21,22] can be approximately described by the Harrison-Perkins-Scott (HPS) tribimaximal mixing (TBM) matrix [23]

$$U_{\text{PMNS}} \approx U_{\text{HPS}} = \begin{pmatrix} \sqrt{\frac{2}{3}} & \frac{1}{\sqrt{3}} & 0 \\ -\frac{1}{\sqrt{6}} & \frac{1}{\sqrt{3}} & \frac{1}{\sqrt{2}} \\ \frac{1}{\sqrt{6}} & -\frac{1}{\sqrt{3}} & \frac{1}{\sqrt{2}} \end{pmatrix}. \qquad (1.1)$$

In U_{HPS}, the solar angle θ_{12} and the atmospheric angle θ_{23} are given by $\theta_{12} = \arcsin(1/\sqrt{3})$ and $\theta_{23} = \pi/4$, whereas the reactor angle θ_{13} vanishes. TBM in the lepton sector, might be regarded as the limit of an underlying flavor symmetry. A flavor symmetry acts on the particle generations and consequently, may be able to predict particle masses and mixings. The actually observed low energy leptonic mixing angles in U_{PMNS} may then be deviations from TBM [24,25] and called nearly tribimaximal lepton mixing [26]. These corrections might stem for instance from higher-dimensional Higgs representations of the underlying theory which generates tribimaximal mixing or by the fact that the theory does not exactly reproduce the matrix of Eq. (1.1). In contrast, the Cabibbo-Kobayashi-Maskawa (CKM) quark mixing matrix V_{CKM} [27,28] is nearly the unit matrix.

In the literature, many models with non-Abelian flavor symmetries have been proposed in order to obtain TBM (*e.g.*, for models based on D_5 – the smallest group with two irreducible two-dimensional representations – see [29, 30], for early models on A_4 – the simplest group with a 3 dimensional representation – and examples using the double covering group of A_4,

---

[1]Meanwhile, the indirect observation of dark matter also confirmed the existence of physics beyond the SM.

see Refs. [31–33] and [34–37]). However, these models generally have difficulties to predict the observed fermion mass hierarchies, the CKM matrix, and have usually a complicated scalar sector (for a discussion see Ref. [38]).

On the other hand, maybe the answer lies in the difference of quark and lepton mixings. In GUTs, quarks and leptons are unified, *i.e.*, they are accommodated in the same multiplets, which was one of the reasons why small lepton mixings were expected in the beginning. However, what if quarks and leptons do not act equal but complementary? This connection between the quark and the lepton sector is implied by the idea of quark-lepton complementarity (QLC) [39–41], which is motivated by the phenomenological observation that the measured solar mixing angle satisfies the relation $\theta_{12} + \theta_C \approx \pi/4$, where $\theta_C \simeq 0.2$ is the Cabibbo angle.

We propose a generalization of QLC, called "extended quark-lepton complementarity" (EQLC) and we systematically construct lepton mass matrices exhibiting nearly tribimaximal neutrino mixing and a small reactor angle $\theta_{13} \approx 0$. The neutrino masses become small due to the type-I seesaw mechanism and large mixings can arise in the charged lepton as well as in the neutrino sector, and in the neutrino sector from Dirac and Majorana masses. The matrix elements of these textures[2] are in the flavor basis expressed by powers of ϵ, which serves as a single small expansion parameter of the matrices. This suggests a model building interpretation of the textures in terms of flavor symmetries, *e.g.*, via the Froggatt-Nielsen mechanism (FN) [48].

We use direct products of cyclic groups to predict via the FN lepton flavor models that provide an excellent fit to current neutrino data with a very small reactor angle. In addition, the Higgs sector is very simple and no fine-tuning of vacuum expectation values (VEVs) is necessary. We also include the quark sector within a SUSY $SU(5)$ GUT scenario since this is often a discriminator of viable models. Thereby, we present a geometrical interpretation of flavor symmetries in an extra dimensional setup and extend the flavor symmetries to non-Abelian ones.

This broad class of non-trivial lepton mass matrix textures are in perfect agreement with neutrino data and we may wonder if there is a possibility to differentiate between them experimentally. Therefore, we survey the lepton flavor violation (LFV) decay rates $Br(\mu \to e\gamma)$, $Br(\tau \to \mu\gamma)$, and $Br(\tau \to e\gamma)$ for the LHC relevant scenario SPS1a' in minimal supergravity at the Lagrangian level. Moreover, we study the branching ratios for the most general CP violating forms of the textures.

---

[2]A texture is a mass matrix with hierarchical entries [42–47].

This work is structured as follows: In Chap. 2 we give a brief overview of the experimental status and present some theoretical approaches aiming to explain the phenomenology. Since flavor symmetries are promising candidates, we introduce them in Chap. 3. An exemplary lepton flavor model is discussed in Chap. 4, where also the notation used throughout this book is introduced. Motivated by this model, we present in Chap. 5 our hypothesis of extended quark-lepton complementarity and generate, based on it, systematically the largest available list of viable CP conserving lepton mass matrices. For these textures we develop in Chap. 6 a method to predict them by flavor symmetries and perform a systematic group space scan. A generalization to CP violating textures and a possible model differentiation by lepton flavor violating rare decays will be analyzed in Chap. 7. The flavor models will in Chap. 8 be extended to SUSY $SU(5)$ GUTs and a geometrical interpretation of the flavor symmetries in 5D will be given. In Chap. 9, we summarize and give a short outlook.

Chapter 2

Status Quo – Experiments and Theories

Experiments provide us with profound knowledge about elementary particles. Recently, it has been discovered that neutrinos which are predicted to be massless in the SM, are massive. The consequence is lepton mixing. However, the quark and lepton sectors are quiet different and various theoretical approaches address these and further questions, e.g., theories using texture zeros, flavor symmetries, SUSY, GUTs, or extra dimensions. In this chapter, we give a brief overview of the experimental status and some theoretical approaches.

2.1 Phenomenological Status

Lepton Sector

One of the most striking properties of elementary particles is their mass. However, the origin of mass in the SM is supposed to be the Higgs mechanism predicting a Higgs boson which is, so far, unobserved. Through spontaneous symmetry breaking of the Higgs field, the elementary particles would acquire mass. The masses of the charged leptons at the mass scale $\mu = M_Z$, are very precisely known [49]

$$\begin{aligned} m_e &= 0.48684727 \pm 0.00000014 \text{ MeV} \;, \\ m_\mu &= 102.75138 \pm 0.00033 \text{ MeV} \;, \\ m_\tau &= 1746.69^{+0.30}_{-0.27} \text{ MeV} \;. \end{aligned} \quad (2.1)$$

For neutrinos we measure only the mass-squared differences

$$\Delta m_{21}^2 = (7.9 \pm 0.3) \cdot 10^{-5} \text{ eV}^2 , \qquad |\Delta m_{31}^2| = (2.5^{+0.2}_{-0.25}) \cdot 10^{-3} \text{ eV}^2 , \qquad (2.2)$$

where $\Delta m_{ij}^2 = m_i^2 - m_j^2$. However, we do not know the absolute mass scale. This can be written as

$$\Delta m_\odot^2 : \Delta m_{\text{atm}}^2 \sim \epsilon^2 , \qquad (2.3)$$

where $\epsilon \simeq 0.2$. Two scenarios are compatible with these data: The so-called normal ordering (NO) and the inverted ordering (IO). There is the possibility that the lightest neutrino is massless. In this limit, i.e., $m_1 = 0$ for NO and $m_3 = 0$ for IO these two scenarios will be called normal (NH) and inverted hierarchy (IH).[1] This is illustrated in Fig. 2.1 (without LSND data [50]). If the neutrino masses are much larger than their mass squared difference Δm^2,

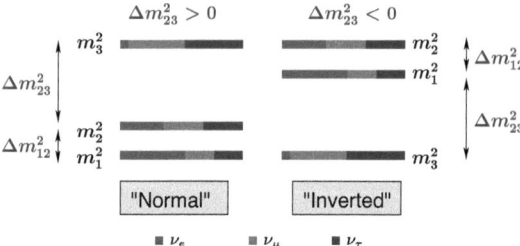

Figure 2.1: Possible neutrino mass orderings.

they are called quasi-degenerate. By using Eq. (2.3) this translates into

$$m_1 : m_2 : m_3 = \epsilon^2 : \epsilon : 1, \quad m_1 : m_2 : m_3 = 1 : 1 : \epsilon, \quad m_1 : m_2 : m_3 = 1 : 1 : 1 . \qquad (2.4)$$

---

[1]Note, in literature, NH and IH is often used instead of NO and IO.

2.1 Phenomenological Status

The mixing of lepton mass and flavor eigenstates is described by the Pontecorvo-Maki-Nakagawa-Sakata matrix

$$U_{\text{PMNS}} = \underbrace{\begin{pmatrix} 1 & 0 & 0 \\ 0 & c_{23} & s_{23} \\ 0 & -s_{23} & c_{23} \end{pmatrix}}_{\text{atmospheric angle}} \underbrace{\begin{pmatrix} c_{13} & 0 & s_{13}e^{-i\delta} \\ 0 & 1 & 0 \\ -s_{13}e^{i\delta} & 0 & c_{13} \end{pmatrix}}_{\text{reactor angle and Dirac } CP \text{ phase}} \underbrace{\begin{pmatrix} c_{12} & s_{12} & 0 \\ -s_{12} & c_{12} & 0 \\ 0 & 0 & 1 \end{pmatrix}}_{\text{solar angle}} \underbrace{\text{diag}(e^{i\phi_1}, e^{i\phi_2}, 1)}_{\text{Majorana phases}}$$

$$= \begin{pmatrix} c_{12}c_{13} & s_{12}c_{13} & s_{13}e^{-i\delta} \\ -s_{12}c_{23} - c_{12}s_{23}s_{13}e^{i\delta} & c_{12}c_{23} - s_{12}s_{23}s_{13}e^{i\delta} & s_{23}c_{13} \\ s_{12}s_{23} - c_{12}c_{23}s_{13}e^{i\delta} & -c_{12}s_{23} - s_{12}c_{23}s_{13}e^{i\delta} & c_{23}c_{13} \end{pmatrix} \text{diag}(e^{i\phi_1}, e^{i\phi_2}, 1),$$
(2.5)

where $c_{ij} = \cos\theta_{ij}$, $s_{ij} = \sin\theta_{ij}$, δ is the Dirac CP violation phase, ϕ_1 and ϕ_2 are two possible Majorana CP violation phases with $\theta_{ij} \in [0, \frac{\pi}{2}]$, and $\delta, \phi_1, \phi_2 \in [0, \pi]$. The values of the currently known mixing parameters are at 3σ (see Ref. [51] for a global fit)

$$\sin^2\theta_{12} = 0.24 \ldots 0.40, \quad |U_{e3}|^2 = |s_{13}e^{-i\delta}|^2 \leq 0.041, \quad \sin^2\theta_{23} = 0.34 \ldots 0.68. \tag{2.6}$$

The Majorana phases are, up to now, unconstrained and are vanishing if neutrinos are not their own anti-particles.

Quark Sector

The quark masses at the mass scale $\mu = M_Z$ [52] are

$$\begin{aligned} m_u &= 1.7 \pm 0.4 \text{ MeV}, & m_c &= 0.62 \pm 0.03 \text{ GeV}, & m_t &= 171 \pm 3 \text{ GeV}, \\ m_d &= 3 \pm 0.6 \text{ MeV}, & m_s &= 54 \pm 8 \text{ MeV}, & m_b &= 2.87 \pm 0.03 \text{ GeV}. \end{aligned} \tag{2.7}$$

Equivalently, in the lepton sector, we have mixings between mass and flavor eigenstates in the quark sector, described by the Cabibbo-Kobayashi-Maskawa (CKM) matrix, where we use the same parameterization as for the PMNS matrix in Eq. (2.5) but without Majorana phases. The experimentally determined limits are

$$|V_{\text{CKM}}| = \begin{pmatrix} 0.97360\ldots 0.97407 & 0.2262\ldots 0.2282 & 0.00387\ldots 0.00405 \\ 0.2261\ldots 0.2281 & 0.97272\ldots 0.97320 & 0.04141\ldots 0.04231 \\ 0.00750\ldots 0.00864 & 0.004083\ldots 0.04173 & 0.999996\ldots 0.999134 \end{pmatrix}, \quad (2.8)$$

this can be expressed in the Wolfenstein parameterization as

$$V_{\text{CKM}} = \begin{pmatrix} 1 - \tfrac{1}{2}\epsilon^2 & \epsilon & A(\rho - i\eta)\epsilon^3 \\ -\epsilon & 1 - \tfrac{1}{2}\epsilon^2 & A\epsilon^2 \\ A(1 - \rho - i\eta)\epsilon^3 & -A\epsilon^2 & 1 \end{pmatrix}, \quad (2.9)$$

where ϵ is of the order the Cabibbo angle $\theta_C \simeq 0.2$, and A, ρ, and η, are order one parameters (for an update see Ref. [53]). This corresponds to the following three angles and Dirac CP phase

$$\begin{aligned} s_{12} &= 0.2243 \pm 0.0016\,, \\ s_{23} &= 0.0413 \pm 0.0015\,, \\ s_{13} &= 0.0037 \pm 0.0005\,, \\ \delta &= 60^\circ \pm 14^\circ\,, \end{aligned} \quad (2.10)$$

where θ_{12} is the so-called Cabibbo angle. These mixing angles are very small compared to the leptonic mixings of Eq. (2.6), where two large mixing angles are present. To find an explanation for this different behavior of quarks and leptons is one of the major challenges theories have to accomplish.

2.2 Theoretical Approaches

In this section, we want to discuss some theoretical approaches trying to explain or reproduce the lepton and quark masses as well as their mixings. This can be achieved by various approaches. One ansatz are so-called texture zeros. Thereby, zero entries in the mass matrices are assumed in order to reduce the number of free parameter and to enhance the predictivity of a texture. In addition, often a certain flavor basis is assumed such as a diagonal charged lepton mass matrix and a diagonal Majorana neutrino mass matrix which is a priori not motivated, for example, by a flavor symmetry. This, of course, neglects a possible origin of mixings stemming

2.2 Theoretical Approaches

from these sectors. A change of the basis or the running of masses from high-scale to low-scale will usually not preserve these texture zeros. Nevertheless, such theories are extensively studied and some can even correlate quark masses and their mixings [54–59] as

$$\left|\frac{V_{ub}}{V_{cb}}\right| = \sqrt{\frac{m_u}{m_c}} \quad \text{and} \quad \left|\frac{V_{td}}{V_{ts}}\right| = \sqrt{\frac{m_d}{m_s}} . \tag{2.11}$$

This arbitrariness disappears as soon as a certain flavor basis is preferred such as in models with flavor symmetries. One possibility are FN models (see Chap. 4) which have usually a simple scalar sector than other flavor models. Such appealing approaches using flavor symmetries will be discussed in more detail in the next chapters.

A further way to engross the thoughts is given by grand unification. Thereby, the SM particles will be accommodated in multiplets of GUT gauge groups, thus having the same properties under the GUT and flavor group. A prominent example is $SU(5)$, where all the SM fermion are accommodated in the multiplets **10**, **5̄**, usually supplemented by a neutrino singlet. Another frequently discussed GUT is $SO(10)$, where all particles are accommodated in the 16-dimensional representation. The attractiveness of GUTs lies in their restrictiveness, i.e., in their predictivity on particles and their properties. In $SU(5)$, for example, the Higgs representation **45** can explain the GUT relations [60]

$$m_b = m_\tau , \quad m_\mu = 3\, m_s , \text{ and } \quad m_d = 3\, m_e . \tag{2.12}$$

Moreover, the smallness of neutrino masses $\sim 10^{-2} \ldots 10^{-1}$ eV can be naturally explained by GUTs [6–8] via the seesaw mechanism [9–17]. In the type-I seesaw, the absolute neutrino mass scale becomes suppressed by the mass scale of heavy SM singlets being at the B-L breaking scale $M_{\text{B-L}} \approx 10^{14}$ GeV which is close to the GUT scale $M_{\text{GUT}} \approx 2 \times 10^{16}$ GeV [18–20]. These SM singlets, i.e., right-handed neutrinos, are predicted by most GUTs. The type-I seesaw mechanism explains therefore not only the smallness of neutrino mass but connects also high energy physics with low energy experiments. There also exist a type-II and meanwhile also a type-III seesaw. These emerge by adding scalar $SU(2)_L$ triplets and fermion triplets, respectively. GUTs have even a further benefit: gauge coupling unification. However, a theory such as SUSY, which might be combined with a GUT, is also able to produce gauge coupling unification and it has even further advantages, e.g., SUSY provides a dark matter candidate. In the era of the large hadron collider (LHC), SUSY is a widely discussed theory since it would be in the discovery range.

However, also more exotic extensions of the SM are possible, such as models with spatial extra dimension. In such models the localization or the propagation profile of the matter fields can explain the mass hierarchy among elementary particles and they can also provide a dark matter candidate.

2.3 Summary

The experimental effort has provided profound knowledge about the world of elementary particles: quark and charged lepton masses as well as quark mixings are measured very accurately. Recently, it has also been discovered that neutrinos are massive, and consequently, that leptons can mix. Indeed, their mixings are even large unlike in the quark sector. However, the absolute mass scale of the neutrinos remains unknown, as well as why the quark and lepton sectors are so different.

Some theoretical approaches address these and further questions. One possibility is to assumes texture zeros in mass matrices which are usually not stable and demanding for justification. This can be done by flavor symmetries which are not only able to predict hierarchies among matrix elements but also are able to relate them. This enables flavor symmetries to predict particle masses and mixings.

The advantages of flavor symmetries can also be combined with the benefits of other theories such as SUSY, GUTs and extra dimensions. Most GUTs for example, suggest the type-I seesaw mechanism which naturally explains the smallness of neutrino mass and connects also high energy physics with low energy experiments. This makes flavor symmetries a powerful and flexible tool in model building.

Chapter

3

Flavor Symmetries

A flavor-, family-, generation- or horizontal-symmetry G_f is a symmetry which acts in generation space and is usually broken at high energies. It commutes usually with the gauge groups and therefore, the transformation properties of fermions under G_f are equal under the gauge groups. Gauge bosons transform only trivial under G_f and a flavor symmetry can therefore be assumed in addition to a GUT[1], *e.g.*, $SU(5) \times S(3)$. The maximal possible flavor symmetry which is compatible with the SM gauge group, is $U(3)^5$ and for the SM with right-handed neutrinos, it is $U(3)^6$. Since gauge interactions are flavor-blind, all particles are invariant under a flavor symmetry $U(3)$. This invariance can be interpreted as a permutation symmetry among the three SM families, *i.e.*, invariance under a S_3 symmetry. First papers considering permutation symmetries are [61–68] or for a review [42]. Extensions of the SM are more restrictive on flavor groups. In $SO(10)$ GUTs, for example, all particles are accommodated in a 16-dimensional representation, and hence the maximal flavor symmetry is $U(3)$.

In the SM, left- and right-handed particles as well as quark and lepton generations can transform differently under G_f or in the same way, and can form a reducible or irreducible representation. From the point of view of GUTs it would be desirable that all generations are unified in one three dimensional representation. However, the top quark transforming as a singlet under G_f and the doublet containing the up- and charm-quark can serve as an

---

[1]Note, for theories which should be valid up to the Planck scale, the origin of the flavor symmetry is assumed to be a continuous gauge symmetry, in order to avoid breaking by gravitational quantum corrections.

explanation for the heavy top-quark mass. In SUSY models the partial unification of the first two generations is used to suppress flavor changing neutral currents (FCNC) in the sfermion sector of these generations. For a solution to the SUSY flavor problem by a flavor symmetry see for example [69].

3.1 A Primer to Flavor Symmetries

In general, before we can think about the accommodation of particles we have to choose a flavor symmetry. However, a flavor symmetry can be continuous or discrete, Abelian or non-Abelian, global or local and can be broken in different ways. Therefore, we briefly discuss these possibilities [70].

Continuous or Discrete Flavor Symmetries?

Flavor symmetries can be continuous such as the gauge groups in the SM or discrete such as crystal symmetries. The advantage of a discrete flavor symmetry is that it has a finite number of representations which dimensions are usually smaller than four. The reason is that we expect to unify just three generations. In addition, no further Goldstone bosons or gauge bosons arise contrary to continuous symmetries. So our choice here drops to a discrete symmetry. However, the origin of the discrete flavor symmetry is assumed to be a continuous gauge symmetry, in order to avoid breaking by gravitational quantum corrections.

Abelian or non-Abelian Flavor Symmetries?

An advantage of discrete non-Abelian flavor symmetries is that most of them have several two or three dimensional representations, what gives us more freedom for the particle assignment and consequently to predict or fit experimental data, respectively. However, Abelian flavor symmetries have generally the merit that they need only a very simple scalar sector to achieve the necessary flavor symmetry breaking.

Local or Global Flavor Symmetries?

The question whether a flavor symmetry should be local or global is equivalent to the one whether we should gauge G_f or not. In order to gauge a flavor symmetry we assume at

3.2 Discrete Flavor Symmetries

a high scale, such as the GUT scale, an anomaly-free continuous group which will be broken spontaneously to our residual discrete flavor group. This continuous group can be gauged in the conventional way. The drawback might be that heavy particles have to be introduced to cancel possible anomalies stemming from the known particles and their representations respectively. For more details on gauged discrete symmetries see an example model based on the double tetrahedral flavor symmetry T' in Ref. [35].

Breaking a Flavor Symmetry

In principle, a flavor symmetry can be broken spontaneously or explicitly. Spontaneous symmetry breaking (SSB) is based on the existence of at least one Higgs boson transforming non-trivially under G_f and acquiring a vacuum expectation value. Models using this mechanism are usually multi-Higgs models and therefore plagued by associated problems such as flavor changing neutral currents (FCNCs) and LFV. Both, FCNCs and LFV, are strongly bounded by experiments, and after SSB, the Higgs potential obey often an accidental symmetry which yields additional Goldstone bosons, which is in conflict with experiments.

3.2 Discrete Flavor Symmetries

General Remarks on Discrete Groups

One of the main fields in which discrete groups are used is solid state physics. There, crystallographic point groups describe symmetries of crystals, and in chemistry they describe the symmetries of atoms and molecules. A point group is a symmetry group which leaves at least one point unmoved. The requirement in crystallography, that this symmetry is present on a lattice requires that only 1, 2, 3, 4, and 6-fold symmetry axes are possible. This restriction is the explanation for the existence of just 32 crystallographic point groups. In general, a discrete group is a group with a discrete topology. In practice, discrete groups often arise as discrete subgroups of continuous Lie groups acting on a geometric space. But they also appear naturally as symmetries of discrete structures (*e.g.*, graphs, tilings, lattices), fundamental groups of topological spaces and so on. In the following sections we want to introduce some of the discrete groups which have already been used as flavor symmetries as well as in other contexts.

Example for Groups

Alternating Groups A_n

The alternating group A_n contains all even permutations of n elements, where even means that the number of performed permutation is even.[2] As an example we show an element of A_4, which is constructed through two permutation:

$$\left[\begin{pmatrix} 1234 \\ 1234 \end{pmatrix} \xrightarrow{1 \to 3} \begin{pmatrix} 1234 \\ 3214 \end{pmatrix} \xrightarrow{2 \to 3} \right] \begin{pmatrix} 1234 \\ 2314 \end{pmatrix} . \tag{3.1}$$

Its group order, $i.e.$, the number of group elements, is $\frac{n!}{2}$. Therefore, it is a subgroup of the symmetric group S_n.

Tetrahedral Group T_d

The tetrahedral group T_d is the symmetry group of the tetrahedron supplemented with the inversion operation and has order 24. It is isomorphic to the group $A_4 \times Z_2$ and is therefore one of the 12 non-Abelian groups of order 24. The pure rotational subgroup of T_d is denoted as T, which is isomorphic to A_4 and has order 12.

Cyclic Groups Z_n

A cyclic group is denoted as Z_n or C_n. It is generated by a single group generator and is Abelian. The generator A satisfies the relation

$$A^n = \mathbb{1} , \tag{3.2}$$

where $\mathbb{1}$ is the identity. If we take the example

$$\begin{pmatrix} 12345678 \\ 23154768 \end{pmatrix} . \tag{3.3a}$$

---

[2]Since the product of two even (or odd) permutations is even, as well as the product of an even and an odd one, the odd permutation of degree n cannot form a group.

3.2 Discrete Flavor Symmetries

we see that $1 \to 2$, $2 \to 3$ and 3 goes again into 1, so they are forming a "cycle". If we now write this as $(123) \equiv$ "$(1 \to 2 \to 3 \to 1)$", we can rewrite Eq. (3.3a) as

$$(123)(45)(67)(8) \ . \tag{3.3b}$$

It consists of four cycles, whereas the cycle (8) is trivial because it only contains the element 8. For every order $n \geq 2$ exists a unique cyclic group. Therefore, cyclic groups of the same order are always isomorphic. Furthermore, subgroups of cyclic groups and all groups of prime order are again cyclic. The cyclic groups of order one or a prime are the only simple Abelian groups. Every Abelian group can be written as a direct product of cyclic subgroups, by computing the characteristic factors. In fact, based on this, we will make use of the factorization of flavor symmetries in semi-direct products ($\ltimes$) of cyclic groups (if applicable) such as

$$\begin{aligned} A_4 &\sim Z_3 \ltimes (Z_2 \times Z_2) \ , \\ T' &\sim Z_2 \ltimes Q \ , \\ \Delta(3\,n^2) &\sim Z_3 \ltimes (Z_n \times Z_n) \ , \end{aligned} \tag{3.4}$$

where Q is the quaternion group. An explicit example often considered in literature is the μ-τ symmetry, i.e., the exchange symmetry of the $2nd$ and $3rd$ lepton generation and assuming a diagonal charged lepton mass matrix. For the neutrinos we introduce the μ-τ symmetric mass matrix form

$$M_\nu = \begin{pmatrix} A & B & B \\ B & C & D \\ B & D & C \end{pmatrix} , \tag{3.5}$$

where $A - D$ are primarily used to fit the observed masses. This is due to the fact that diagonalization of M_ν leads, independently of $A - D$, to a maximal θ_{23} and a vanishing θ_{13} mixing angle of the PMNS matrix in agreement with experiments.

Double Groups

The double groups descend from the point groups by adding the operation R which has the matrix representation $\pm \mathbb{1}_{n \times n}$ for a n dimensional representation. Their elements are called single-valued if the matrix representation of R is $+\mathbb{1}_{n \times n}$ otherwise double-valued. The order of the double group is twice the one of the originale group and the groups are denoted with a

', e.g., the double group of T_n is T'_n. Since the single-valued representations are the same as for the single groups, the double-valued representations are new. For Abelian single groups the corresponding double group can, but do not have to be, non-Abelian. For the cyclic groups Z_n the double groups are isomorphic to Z_{2n} and thus they are Abelian.

3.3 Summary

Flavor symmetries G_f commute usually with the gauge groups, e.g., $SU(5) \times S(3)$ and gauge bosons transform only trivially under G_f. Gauge interactions are flavor-blind what translates into an invariance of a $U(3)$ symmetry. In the SM, the maximal possible flavor symmetry is $U(3)^5$ and $U(3)^6$ with right-handed neutrinos, respectively.

Abelian flavor symmetries have generally the merit that they need only a simple scalar sector. However, as the origin of a discrete flavor symmetry, an anomaly-free continuous gauge symmetry shall be assumed, which is broken spontaneously to the residual discrete group. This avoids breaking by gravitational quantum corrections. The drawback can be that heavy particles might have to be introduced to cancel possible anomalies stemming from the known particles. Flavor models with SSB are usually multi-Higgs models and therefore often show problems associated with FCNCs, LFV, and/or Goldstone bosons.

Every Abelian group can be written as a direct product of cyclic subgroups.

$$A_4 \sim Z_3 \ltimes (Z_2 \times Z_2),$$
$$T' \sim Z_2 \ltimes Q,$$
$$\Delta(3n^2) \sim Z_3 \ltimes (Z_n \times Z_n).$$

Therefore, we can use isomorphisms among groups to express some discrete groups as semidirect products of cyclic groups.

Chapter 4

Flavored Models

In this chapter, we present the notation and flavor model setup we will use throughout this book. For this we introduce the Froggatt-Nielsen mechanism leading to a simple scalar sector and a single expansion parameter $\epsilon \simeq 0.2$ which is of the order the Cabibbo angle. For illustration, we present a $Z_5 \times Z_9$ lepton flavor model and new sum rules.

4.1 Froggatt-Nielsen Mechanism

The idea that Higgs fields break a gauge symmetry such that quark and lepton masses arise from higher-dimensional terms was proposed by Froggatt and Nielsen [48] and is illustrated in Fig. 4.1. The left- and right-handed SM fermions ψ_L and ψ_R obtain masses via couplings to

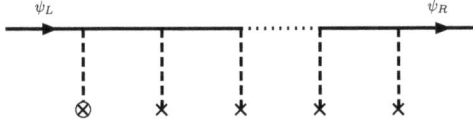

Figure 4.1: Froggatt-Nielsen mechanism generating effective SM fermion (ψ_L and ψ_R) masses which are suppressed by a factor $\epsilon^n = (v/M_F)^n$. Thereby, v is a universal flavon VEV (crosses), M_F the universal mass of superheavy fermions (solid lines), and the SM Higgs VEV is denoted by a encircled cross.

the SM Higgs and in addition to superheavy fermions with universal mass M_F transforming non-trivially under G_f, and charged SM singlet scalars breaking G_f by acquiring a universal

VEV. This leads to effective SM fermion masses that are suppressed by a factor $\epsilon^n = (v/M_F)^n$, where n is determined by the flavor quantum numbers of ψ_L and ψ_R. A model based on this mechanism will be presented and discussed in more detail in the next section.

4.2 A $Z_5 \times Z_9$ Lepton Flavor Model

In this section we present the notation which we will use throughout the following chapters as well as a lepton flavor model based on $G_f = Z_5 \times Z_9$. This particular group turns out to be very fruitful for constructing lepton flavor models (cf., Fig. 6.1). All model building tools used in this section will furthermore be important for the following chapters.

4.2.1 Notation and Model Outline

This model is based on the FN mechanism, i.e., the flavons have charge ± 1 and the effective mass terms are suppressed by a power n of the factor $\epsilon = (v/M_F)$ which is assumed to be of the order the Cabibbo angle, i.e., $\epsilon \simeq 0.2$. The power n of the suppression is solely determined by the fermion quantum numbers.

We will use ϵ as a single expansion parameter. Therefore, we parameterize and express also the fermion mass ratios and mixings by ϵ. Even though we present here a lepton flavor model, we will give the equivalent relations for quarks since we will use them later on in Chap. 8. Barring numerical factors, the fermion mass ratios we are using for charged leptons[1], effective neutrinos, up-type quarks and down-type quarks are

$$\begin{aligned}
m_e : m_\mu : m_\tau &= \epsilon^4 : \epsilon^2 : 1 \,, \\
m_1 : m_2 : m_3 &= \epsilon^2 : \epsilon : 1 \,, \\
m_u : m_c : m_t &= \epsilon^6 : \epsilon^4 : 1 \,, \\
m_d : m_s : m_b &= \epsilon^4 : \epsilon^2 : 1 \,.
\end{aligned} \quad (4.1)$$

Note, these mass ratios have all to be understood as order of magnitude relations and depend on the energy scale. In addition, we propose to write the mass eigenvalues of M_D and M_R also

---

[1] The particular choice for the charged lepton mass ratios is motivated by an $SU(5)$ GUT compatibility, where $M_e = M_d^T$. However, a modification, e.g., the implementation of the Georgi-Jarlskog relation $m_\mu : m_\tau = 3m_s : m_b$ would not change any of our results for the textures.

4.2 A $Z_5 \times Z_9$ Lepton Flavor Model

as powers of ϵ. They read

$$m_1^D : m_2^D : m_3^D = \epsilon^a : \epsilon^b : \epsilon^c \quad \text{and} \quad m_1^R : m_2^R : m_3^R = \epsilon^{a'} : \epsilon^{b'} : \epsilon^{c'}, \tag{4.2}$$

where a, b, c, a', b', and c', are suitable non-negative integers, and we define the absolute mass scales by $m_3^D = m_D \epsilon^c$ and $m_3^R = M_{B-L}\epsilon^{c'}$. In addition, we use our freedom to order the masses of M_R strictly hierarchical and choose $a' \leq b' \leq c'$ (alternatively we could order M_D). For completeness, we want to parameterize here also the TBM form of the PMNS matrix and the CKM mixings in terms of maximal mixing and powers of ϵ.[2] Then, the lepton mixing angles read

$$\theta_{12} = \tfrac{\pi}{4} - \epsilon \;, \quad \theta_{13} = 0 \;, \quad \theta_{23} = \tfrac{\pi}{4} \;, \tag{4.3}$$

and the CKM mixings for quarks are

$$V_{us} = \epsilon \;, \quad V_{cb} = \epsilon^2 \;, \quad V_{ub} = \epsilon^3 \;. \tag{4.4}$$

The relation for the leptonic mixing angle θ_{12} is called quark-lepton-complementarity (QLC) since it relates quarks and leptons. We will discuss and extend this relation in Sec. 5.1.1.

The lepton Yukawa couplings and mass terms implementing the type-I seesaw mechanism read

$$\mathcal{L}_Y = -(Y_\ell)_{ij} H^* \ell_i e_j^c - (Y_D)_{ij} i\sigma^2 H \ell_i \nu_j^c - \frac{1}{2}(M_R)_{ij}\nu_i^c \nu_j^c + \text{H.c.}, \tag{4.5}$$

where ℓ_i, e_i^c, and ν_i^c, are the left-handed leptons, the right-handed charged leptons, and the right-handed neutrinos, and $i = 1, 2, 3$ is the generation index. H is the SM Higgs doublet, Y_ℓ and Y_D are the Dirac Yukawa coupling matrices of the charged leptons and neutrinos, and M_R is the Majorana mass matrix of the right-handed neutrinos with entries of the order the $B-L$ breaking scale $M_{B-L} \sim 10^{14}$ GeV. After electroweak symmetry breaking, H develops a vacuum expectation value $\langle H \rangle \sim 10^2$ GeV, and the mass terms of the leptons become

$$\mathcal{L}_{\text{mass}} = -(M_\ell)_{ij} e_i e_j^c - (M_D)_{ij}\nu_i \nu_j^c - \frac{1}{2}(M_R)_{ij}\nu_i^c \nu_j^c + \text{H.c.}, \tag{4.6}$$

where $M_\ell = \langle H \rangle Y_\ell$ is the charged lepton and $M_D = \langle H \rangle Y_D \sim 10^2$ GeV the Dirac neutrino mass matrix. After integrating out the right-handed neutrinos, the seesaw mechanism leads to an

---

[2] However note that our $Z_5 \times Z_9$ lepton model actually predicts explicit PMNS mixing angles.

effective Majorana neutrino mass matrix

$$M_{\text{eff}} = -M_D M_R^{-1} M_D^T , \qquad (4.7)$$

with entries of the order 10^{-2} eV in good agreement with observation.

The leptonic Dirac mass matrices M_ℓ and M_D, and the Majorana mass matrices M_R and M_{eff} can be diagonalized by

$$\begin{aligned} M_\ell &= U_\ell M_\ell^{\text{diag}} U_{\ell'}^\dagger , & M_{\text{eff}} &= U_\nu M_{\text{eff}}^{\text{diag}} U_\nu^T , \\ M_D &= U_D M_D^{\text{diag}} U_{D'}^\dagger , & M_R &= U_R M_R^{\text{diag}} U_R^T , \end{aligned} \qquad (4.8)$$

where $U_\ell, U_{\ell'}, U_D, U_{D'}, U_R,$ and U_ν, are unitary mixing matrices, whereas M_ℓ^{diag}, M_D^{diag}, M_R^{diag}, and $M_{\text{eff}}^{\text{diag}}$ are diagonal mass matrices with positive entries. The mass eigenvalues of the charged leptons and neutrinos are given by $M_\ell^{\text{diag}} = \text{diag}(m_e, m_\mu, m_\tau)$ and $M_{\text{eff}}^{\text{diag}} = \text{diag}(m_1, m_2, m_3)$.

A unitary mixing matrix U_x can always be written as a product of the form

$$U_x = D_x \hat{U}_x K_x, \qquad (4.9)$$

where $\hat{U}_x$ is a CKM-like matrix parameterized according to Eq. (2.5). Here, we follow the conventions and definitions given in Ref. [71]. $D_x = \text{diag}(e^{i\varphi_1^x}, e^{i\varphi_2^x}, e^{i\varphi_3^x})$ and $K_x = \text{diag}(e^{i\alpha_1^x}, e^{i\alpha_2^x}, 1)$ are diagonal phase matrices, where the index $x \in \{\ell, \ell', D, D', R, \nu\}$. The phases in D_x and K_x are all in the range $[0, \pi]$. The PMNS matrix, by taking rephasing invariance into account, can be written as

$$U_{\text{PMNS}} = U_\ell^\dagger U_\nu = \hat{U}_{\text{PMNS}} K_{\text{Maj}}, \qquad (4.10)$$

where $K_{\text{Maj}} = \text{diag}(e^{i\phi_1}, e^{i\phi_2}, 1)$ contains the Majorana phases ϕ_1 and ϕ_2, and U_ℓ and U_ν are in general of the form of Eq. (4.9). The effective neutrino mass matrix of Eq. (4.7) is therefore

$$M_{\text{eff}}^{\text{th}} = -D_D \hat{U}_D \tilde{K} M_D^{\text{diag}} \hat{U}_{D'}^\dagger \tilde{D} \hat{U}_R^* (K_R^*)^2 (M_R^{\text{diag}})^{-1} \hat{U}_R^\dagger \tilde{D} \hat{U}_{D'}^* M_D^{\text{diag}} \tilde{K} \hat{U}_D^T D_D , \qquad (4.11)$$

where we have merged $\tilde{K} = K_D^* K_{D'}$ and $\tilde{D} = D_{D'}^* D_R^*$.[3] Note, we have introduced an extra superscript "th" for "theoretical", since none of the mass and mixing parameters on the right-hand side of Eq. (4.11) are directly measurable in neutrino oscillations. Therefore, we introduce also

---
[3]In the CP conserving case, the matrix $(K_R^*)^2$ drops out of the expression for M_{eff} in Eq. (4.11).

4.2 A $Z_5 \times Z_9$ Lepton Flavor Model

$M_{\text{eff}}^{\text{exp}}$, which involves the matrices $M_{\text{eff}}^{\text{diag}}$ and U_{PMNS} containing the experimentally accessible mass and mixing parameters. For this purpose, we insert Eq. (4.10) in the expression for M_{eff} in Eq. (4.8) and obtain

$$M_{\text{eff}}^{\text{exp}} = D_\ell \widehat{U}_\ell K_\ell \widehat{U}_{\text{PMNS}} K_{\text{Maj}}^2 M_{\text{eff}}^{\text{diag}} \widehat{U}_{\text{PMNS}}^T K_\ell \widehat{U}_\ell^T D_\ell \ . \tag{4.12}$$

Note that $(K_{\text{Maj}})^2$ drops out in the CP conserving case, and that $M_{\text{eff}}^{\text{th}} \simeq M_{\text{eff}}^{\text{exp}} \simeq M_{\text{eff}}$, since $M_{\text{eff}}^{\text{th}}$ and $M_{\text{eff}}^{\text{exp}}$ are just different parameterizations of M_{eff}.

4.2.2 Lepton Masses and Mixings

In our model, we assume the following flavor charges[4] for the right-handed leptons e_i^c, for the left-handed lepton doublets ℓ_i, and for the right-handed neutrinos ν_i^c,

$$\begin{aligned} e_1^c, e_2^c, e_3^c &\sim (3,8), (4,3), (0,3) \ , \\ \ell_1, \ell_2, \ell_3 &\sim (0,4), (3,7), (4,6) \ , \\ \nu_1^c, \nu_2^c, \nu_3^c &\sim (0,8), (2,4), (1,0) \ . \end{aligned} \tag{4.13}$$

Here, the first entry in each bracket refers to the quantum number associated with Z_5 and the second one to Z_9, respectively. This simple setup already determines the mass matrix textures for the leptons and exhibits a very simple scalar sector. Up to an undetermined overall mass scale, the resulting mass matrix textures are

$$M_\ell \sim \epsilon \begin{pmatrix} \epsilon^4 & \epsilon^2 & \epsilon \\ \epsilon^3 & \epsilon^2 & \epsilon^2 \\ \epsilon^5 & \epsilon & 1 \end{pmatrix} \ , \quad M_D \sim \epsilon^2 \begin{pmatrix} \epsilon & \epsilon & \epsilon^3 \\ \epsilon^3 & 1 & \epsilon \\ \epsilon^3 & 1 & \epsilon \end{pmatrix} \ , \quad M_R \sim \epsilon^2 \begin{pmatrix} 1 & \epsilon^3 & 1 \\ \epsilon^3 & 1 & \epsilon^4 \\ 1 & \epsilon^4 & 1 \end{pmatrix} \ . \tag{4.14}$$

These mass matrix textures predict the relative order of magnitude of each matrix element. However, order one couplings are undetermined and have to be fitted. Supplemented with these

---

[4]Remember that this is equivalent to an embedding into representations of flavor symmetries.

Yukawa couplings and using $\epsilon = 0.2$ we find

$$M_\ell \sim \begin{pmatrix} -0.00153685 & 0.0397748 & 0.194647 \\ 0.00795287 & 0.0461743 & 0.0306293 \\ -6.5 \times 10^{-6} & 0.192955 & 0.960086 \end{pmatrix},$$

$$M_D \sim \begin{pmatrix} -0.201901 & -0.134642 & 0.00729065 \\ -0.00108898 & -0.782587 & 0.173945 \\ 0.00159886 & -0.603486 & -0.118982 \end{pmatrix}, \quad (4.15)$$

$$M_R \sim \begin{pmatrix} 0.6 & 0 & -0.4 \\ 0 & 1 & 0 \\ -0.4 & 0 & 0.6 \end{pmatrix}.$$

Note, in Chap. 6, we explain the origin and values of order one couplings (the model is #5 of Table 6.1). Here, we have neglected the overall ϵ suppression factor for each matrix. However, for a better comparison, we have not normalized the textures to the maximal absolute value of each matrix. In Eq. (4.15), we have used only Yukawa couplings which are naturally of order one, *i.e.*, they are in the range of ϵ and $1/\epsilon$, in order not to blur the texture entry predicted by the model. Diagonalization of the mass matrices of Eq. (4.15) leads, besides the lepton mass ratios of Eq. (4.1) to the following Dirac and Majorana masses

$$M_D^{\text{diag}} \sim \text{diag}(\epsilon, 1, \epsilon) \quad \text{and} \quad M_R^{\text{diag}} \sim \text{diag}(\epsilon, 1, 1), \quad (4.16)$$

and mixings (mixing angles and phases are given according to the notation of Sec. 4.2.1)

$$\begin{aligned}
(\theta_{12}^\ell, \theta_{13}^\ell, \theta_{23}^\ell, \delta^\ell, \alpha_1^\ell, \alpha_2^\ell) &\sim (0, \epsilon, \epsilon^2, 0, \pi, 0), \\
(\theta_{12}^{\ell'}, \theta_{13}^{\ell'}, \theta_{23}^{\ell'}, \delta^{\ell'}, \alpha_1^{\ell'}, \alpha_2^{\ell'}) &\sim (\epsilon, 0, \epsilon, 0, 0, 0), \\
(\theta_{12}^D, \theta_{13}^D, \theta_{23}^D, \delta^D, \varphi_1^D, \varphi_2^D, \varphi_3^D) &\sim (\epsilon, \tfrac{\pi}{4}, \tfrac{\pi}{4}, \pi, 0, 0, \pi), \\
(\theta_{12}^{D'}, \theta_{13}^{D'}, \theta_{23}^{D'}, \delta^{D'}, \alpha_1^{D'}, \alpha_2^{D'}) &\sim (\epsilon^2, \tfrac{\pi}{4}, \epsilon^2, 0, \pi, \pi), \\
(\theta_{12}^R, \theta_{13}^R, \theta_{23}^R, \delta^R, \varphi_1^R, \varphi_2^R, \varphi_3^R) &\sim (0, \tfrac{\pi}{4}, 0, 0, 0, 0, \pi).
\end{aligned} \quad (4.17)$$

This means that we have large mixings only in the neutrino sector. However, in the neutrino sector we have maximal mixings in both sectors, in the Dirac as well as in the Majorana sector, and nontrivial charged lepton mixing. The fact that all mixing angles of U_ℓ, U_D, $U_{D'}$, and U_R,

4.2 A $Z_5 \times Z_9$ Lepton Flavor Model

are powers of ϵ and $\pi/4$ looks somehow like a "miracle", but is, in fact, an outcome of our approach which we will introduce in Chap. 5. This method provides also the Yukawa couplings used in Eq. (4.15).

4.2.3 New Sum Rules

So far, we have not discussed the PMNS mixings of our model. An expansion in ϵ up to $\mathcal{O}(\epsilon^3)$ for U_{PMNS} leads to the following new sum rules

$$\theta_{12} = \tfrac{\pi}{4} - \tfrac{\epsilon}{\sqrt{2}} - \tfrac{\epsilon^2}{4}, \qquad \theta_{13} = \mathcal{O}(\epsilon^3), \qquad \theta_{23} = \tfrac{\pi}{4} + \tfrac{\epsilon}{\sqrt{2}} - \tfrac{3}{4}\epsilon^2. \qquad (4.18)$$

Note that the predicted mixing angels deviate from exact TBM shown in Eq. (4.3) as well as from QLC. This nearly tribimaximal mixing makes the model testable in future neutrino oscillation experiments. Especially, the deviation from maximal atmospheric mixing can be established at the 3σ confidence level (CL) by the T2K and NOνA experiments [72] and the sign of the deviation from maximal mixing (the octant) with a neutrino factory at the 3σ CL for $\sin^2 2\theta_{13} \gtrsim 10^{-2.5}$ or at 90% CL otherwise [73].

However, the sum rules of Eq. (4.18) could, in principle, be the result of an accidental fine-tuning of Yukawa couplings. Therefore, we have varied the order one couplings of M_ℓ, M_D, and M_R independently within 1% range. The results for the lepton mixings are shown in Fig. 4.2. There we can see that a 1% variation of Yukawa couplings results in a 1% variation of the

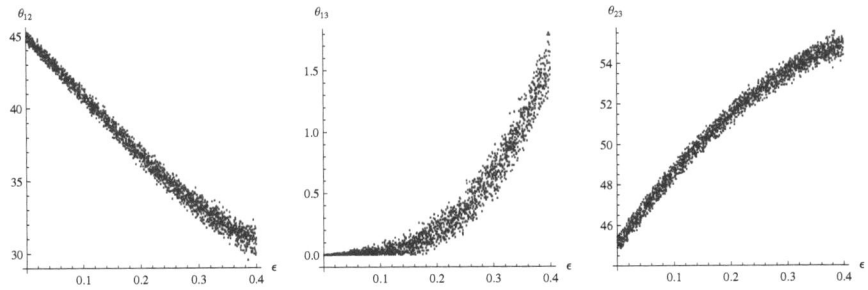

Figure 4.2: Effect of a 1% variation of Yukawa couplings in all mass matrices on the PMNS mixing angles.

mixing angles. This is expected when no fine-tuning is present. We can also see the limit of

$45°$ for θ_{12} and θ_{23} in the case of vanishing ϵ, as well as the exponential behavior of degree three of θ_{13}, which is thus not due to an accidental cancellation of large mixings (for an anlytical discussion of sum rules from textures, see [74]).

4.3 Summary

In this chapter, we have presented our setup for model building with flavor symmetries. The neutrino masses become small due to the type-I seesaw mechanism and we achieve a simple scalar sector via the Froggatt-Nielsen mechanism. Thereby, we are using a single expansion parameter $\epsilon \simeq \theta_C \simeq 0.2$, which is of the order the Cabibbo angle.

This was done for a lepton flavor model based on the flavor symmetry $Z_5 \times Z_9$. On this basis, we have introduced our notation and setup which we will use throughout the following chapters. In addition, we have discussed aspects such as new sum rules predicted by our model. These sum rules make the model testable in future experiments such as T2K, NOνA, and neutrino factories.

Chapter

5

Textures – A Bottom-Up Approach

In this chapter, we introduce our hypothesis of extended quark-lepton complementarity. Based on it, we generate the largest list of type-I seesaw realizations and textures available in literature [75–77].[1] We show that special cases often considered, such as having a symmetric Dirac mass matrix or small mixing among charged leptons, constitute only a tiny fraction of our possibilities. An exemplary list of 72 realizations with corresponding textures is shown. This list represents a broad class of CP conserving realizations in the sense that large mixings can arise in the charged lepton and the neutrino sector. Moreover, in the neutrino sector maximal mixing can occur in the Dirac or in the Majorana sectors. The mass spectrum of M_D and M_R can be hierarchical or semi-hierarchical. All of these mass matrices lead to nearly tribimaximal mixing, $i.e.$, to a very small reactor angle $\theta_{13} \lesssim 1°$ and a deviation from the atmospheric mixing angle testable in future experiments.

---

[1]We discuss IH and DG neutrinos in Ref. [75]. In Ref. [71], we covered realizations and textures based on Majorana neutrinos and the type-II seesaw, respectively.

5.1 Textures with Extended Quark-Lepton Complementarity

5.1.1 Extended Quark-Lepton Complementarity

In Sec. 4.2.1, we already became acquainted with the quark-lepton complementarity

$$\theta_{12} \approx \frac{\pi}{4} - \epsilon \,, \qquad \theta_{23} + \theta_{cb} \approx \frac{\pi}{4} \,. \tag{5.1}$$

QLC can serve as explanation for the deviation of θ_{12} from maximal mixing and has been studied from many different points of view: as a correction to bimaximal mixing [78–80], together with sum rules [81,82], with stress on phenomenological aspects [83,84], in conjunction with parameterizations of U_{PMNS} in terms of θ_C [85–89], with respect to statistical arguments [90], by including renormalization group effects [91,92], and in model building realizations [93–97]. We are now going to extend this idea of connecting the quark and the lepton sector to explain the mixings. In Ref. [71,75], we have suggested a generalization of QLC called "extended QLC" (EQLC). Thereby, the mixing angles of all left- and right-handed fermions $U_\ell, U_D, U_{D'}$, and U_R, can take any of the values $\frac{\pi}{4}, \epsilon, \epsilon^2, ..., 0$. However, unless otherwise noted, we will identify all terms of order ϵ^n with $n \geq 3$ simply by "0" since the current 1σ error on the leptonic mixings is at most of the order ϵ^2 (see, e.g., Ref. [51]). A further motivation to express the mixings in this way is given, besides the observed quark mixing angles given in Eq. (4.4), by a μ-τ flavor symmetry, which naturally leads to maximal mixing, cf., Sec. 3.2. This applies not only to the mixings but also to the unmeasured mass ratios of M_D and M_R in Eq. (4.2), i.e., a, b, c, a', b', and c', are assumed to be 0, 1, or 2. In combination with the fermion mass ratios parameterized by ϵ, cf., Eq. (4.1), this allows us to express all masses and mixings.

However, note that the mixings of Dirac and Majorana neutrinos are present at high energies and the PMNS mixings are measured at low energies. Consequently, EQLC might be realized but is a challenge for experiments. We discuss this in more detail in the next section.

5.1.2 Mass Matrix Production

In Sec. 4.2, we have seen that textures can be predicted by flavor symmetries. However, we will follow a bottom-up approach: we will "search" for viable mass matrix textures without

5.1 Textures with Extended Quark-Lepton Complementarity

motivation – for the moment – by flavor symmetries. Chaps. 6 and 8 are devoted to the systematic search for flavor symmetries. Nevertheless, we do not stick to a certain flavor basis or use invariants since we would lose information about high energy physics. In oscillation experiments, we measure the lepton mixing matrix U_{PMNS}, which is a product of the charged lepton and neutrino mixing matrix $U_\ell^\dagger$ and U_ν. Therefore, it would not be meaningful at all to search for one "viable" mass matrix texture but instead we have to look for a whole texture-set consisting of M_ℓ, M_D, and M_R (in case of the type-I seesaw). This approach is illustrated in Fig. 5.1.

In order to ensure the viability of our texture sets, we use the experimental values such as masses and mixings within their experimental errors. For this we generate all possibilities for the neutrino mass matrix $M_{\text{eff}}^{\text{exp}}$ in Eq. (4.12), where we use the current best-fit values[2] [51] of U_{PMNS}, i.e.,

$$\theta_{12} = \pi/4 - \epsilon, \quad \theta_{13} = 0, \quad \theta_{23} = \pi/4 \ .$$

Note that these values correspond also to TBM as well as to an interesting symmetry limit in certain neutrino mass models [38]. In addition, we assume the normal hierarchical neutrino spectrum of Eq. (4.1) and the hypothesis of EQLC, i.e.,

$$\theta_{ij}^x \in \{\pi/4, \epsilon, \epsilon^2, 0\}, \tag{5.2}$$

with $x \in \{\ell, \ell', D, D', R, \nu\}$. The phases $\delta^x, \varphi_1^x, \varphi_2^x, \varphi_3^x, \alpha_1^x, \alpha_2^x \in \{0, \pi\}$ are here assumed to be CP conserving. This ambiguity of U_ℓ now leads to various neutrino mass matrices $M_{\text{eff}}^{\text{exp}}$, which reproduce the experimental mass and mixing parameters, by construction.

However, so far, we have only a set of viable effective mass matrices $M_{\text{eff}}^{\text{exp}}$ but not the desired sets of textures M_ℓ, M_D, and M_R, valid at high energies. Therefore, we generate, in analogy with $M_{\text{eff}}^{\text{exp}}$, all theoretically possible effective neutrino mass matrices $M_{\text{eff}}^{\text{th}}$ of Eq. (4.11). Thereby, we again assume our hypothesis of EQLC, i.e., all mixings of U_D, $U_{D'}$, and U_R, are taken from the set shown in Eq. (5.2), the phases are assumed to be CP conserving, and the mass ratios[3] of M_D^{diag} and M_R^{diag} can be expressed as powers of ϵ up to order two, since this

---

[2] These values could be experimentally confirmed or rejected in the coming years. See, e.g., Refs. [72, 98] for long-baseline experiments on a scale of the coming ten years, Refs. [99,100] for an up-scale reactor θ_{12} measurement, Ref. [101] for the potential of various different superbeam upgrades, and Ref. [102–104] for a neutrino factory measurement. A different choice for the parameters can be equally well applied, but it will change the final results, e.g., if θ_{13} is indeed found to be non zero.

[3] Note that we will always set the heaviest mass eigenvalue equal to the absolute neutrino mass scale.

corresponds to the experimental errors in the neutrino sector, *cf.*, Eq. (4.2). Note, in the case of CP violation, some textures may change due to cancellations, so the number of textures will increase. Nevertheless, a systematic analysis with phases between 0 and 2π is up to now not possible due to the lack of computing power but this might change in some years (for textures with type-II seesaw mechanism, see Ref. [105]).

Next, we have to confront theory with experiment, *i.e.*, we match all possibilities (parameter combinations) of $M_{\text{eff}}^{\text{th}}$ with all possibilities of $M_{\text{eff}}^{\text{exp}}$ for $\epsilon = 0.2$, since this are just different parameterizations for the same neutrino mass matrix. In total, our procedure requires that we systematically scan more than 20 trillion different possibilities. The matching

$$M_{\text{eff}}^{\text{th}}|_{\epsilon=0.2} \simeq M_{\text{eff}}^{\text{exp}}|_{\epsilon=0.2} \tag{5.3}$$

will be done in accordance with the experimental errors $\mathcal{O}(\epsilon^3)$. The overall neutrino mass scale, *i.e.*, $m_\nu = m_D^2/M_R$ will thereby automatically factored out. This leads to 1981 sets of parameters generating viable mass matrices $\{M_D, M_R, M_\ell\}$, which we will call a (seesaw) realization.[4] These realizations contain all relevant parameters (mass ratios, mixings, and phases) to reconstruct the mass matrices $\{M_D, M_R, M_\ell\}$ which we can analytically expand in powers of ϵ. By taking the leading order term of each matrix element, we obtain the texture entries with their Yukawa couplings. In other words, by expanding the mass matrices in ϵ and by neglecting all but the leading order we obtain the corresponding mass matrix textures.

---

[4]Note that we choose $U_{\ell'} = $ since it does not affect any phenomenology. However, in models, $U_{\ell'}$ is predicted and therefore, we are going to consider it in the next chapter.

5.1 Textures with Extended Quark-Lepton Complementarity

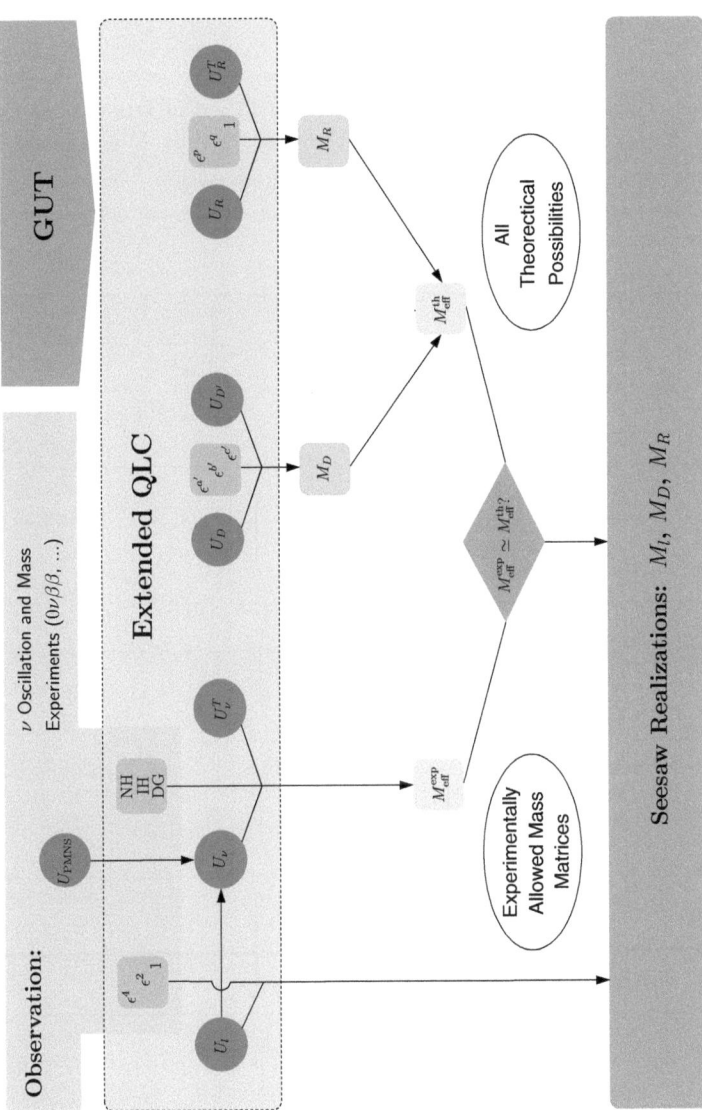

Figure 5.1: Procedure for obtaining the seesaw realizations and texture sets in extended quark-lepton complementarity.

5.2 Essence of EQLC-Mass Matrices

5.2.1 Seesaw Realizations

The 1981 viable sets of seesaw realizations obtained in Sec. 5.1.2 is, of course, too huge to be presented here. Therefore, we show an exemplary list of 72 texture sets with corresponding realizations in Tables 5.1 and A.1 and introduce a parameter $\xi \in \{0, \epsilon^2\}$ which nicely demonstrates the influence of the mixing angles on certain texture entries. The "quality" of these seesaw realizations and their properties such as Yukawa couplings are discussed in the next section.

#	M_ℓ	M_D	M_R	m_i^D/m_D m_i^R/M_{B-L}	$(\theta_{12}^\ell, \theta_{13}^\ell, \theta_{23}^\ell)$ $(\theta_{12}^D, \theta_{13}^D, \theta_{23}^D)$ $(\theta_{12}^{D'}, \theta_{13}^{D'}, \theta_{23}^{D'})$ $(\theta_{12}^R, \theta_{13}^R, \theta_{23}^R)$
1	$\begin{pmatrix} 0 & 0 & \epsilon \\ 0 & \epsilon^2 & 1 \\ 0 & \epsilon^2 & 1 \end{pmatrix}$	$\begin{pmatrix} \epsilon^2 & \epsilon & 0 \\ \xi & 1 & 1 \\ \epsilon^2 & \xi & \epsilon \end{pmatrix}$	$\begin{pmatrix} \epsilon^2 & 0 & 0 \\ 0 & 1 & 1 \\ 0 & 1 & 1 \end{pmatrix}$	$(\epsilon^2, 1, \epsilon)$ $(\epsilon^2, \epsilon, 1)$	$(\xi, \epsilon, \frac{\pi}{4})$ $(\epsilon, \frac{\pi}{4}, \xi)$ $(\xi, \xi, \frac{\pi}{4})$ $(\xi, 0, \frac{\pi}{4})$
2	$\begin{pmatrix} 0 & 0 & 1 \\ 0 & \epsilon^2 & \epsilon \\ 0 & 0 & 1 \end{pmatrix}$	$\begin{pmatrix} \epsilon & 1 & \epsilon \\ \epsilon & 1 & \epsilon^2 \\ \epsilon & 1 & \epsilon \end{pmatrix}$	$\begin{pmatrix} 1 & \epsilon & 1 \\ \epsilon & 1 & \epsilon \\ 1 & \epsilon & 1 \end{pmatrix}$	$(\epsilon, 1, \epsilon)$ $(\epsilon, 1, 1)$	$(\epsilon, \frac{\pi}{4}, \epsilon)$ $(\frac{\pi}{4}, \frac{\pi}{4}, \epsilon^2)$ $(\epsilon^2, \epsilon, 0)$ $(\epsilon^2, \frac{\pi}{4}, \epsilon)$
3	$\begin{pmatrix} 0 & \epsilon^2 & 1 \\ 0 & \epsilon^2 & 1 \\ 0 & \epsilon^2 & 1 \end{pmatrix}$	$\begin{pmatrix} \epsilon & \epsilon^2 & \epsilon \\ \epsilon^2 & 1 & \epsilon \\ \epsilon & \epsilon & \epsilon \end{pmatrix}$	$\begin{pmatrix} 1 & \epsilon & 1 \\ \epsilon & 1 & \epsilon \\ 1 & \epsilon & 1 \end{pmatrix}$	$(\epsilon, 1, \epsilon)$ $(\epsilon, 1, 1)$	$(\frac{\pi}{4}, \frac{\pi}{4}, \frac{\pi}{4})$ $(0, \frac{\pi}{4}, \epsilon)$ $(\epsilon^2, \epsilon^2, \epsilon)$ $(\epsilon, \frac{\pi}{4}, \epsilon^2)$
4	$\begin{pmatrix} 0 & 0 & \epsilon \\ 0 & \epsilon^2 & 1 \\ 0 & \epsilon^2 & 1 \end{pmatrix}$	$\begin{pmatrix} \epsilon & \epsilon & \epsilon \\ \epsilon & \epsilon & \epsilon \\ \xi & 1 & 1 \end{pmatrix}$	$\begin{pmatrix} \epsilon & \epsilon & \epsilon \\ \epsilon & 1 & \epsilon^2 \\ \epsilon & \epsilon^2 & 1 \end{pmatrix}$	$(\epsilon, \epsilon, 1)$ $(\epsilon, 1, 1)$	$(\epsilon, \epsilon, \frac{\pi}{4})$ $(\frac{\pi}{4}, \xi, \xi)$ $(\epsilon, \xi, \frac{\pi}{4})$ $(\epsilon, \epsilon, \xi)$
5	$\begin{pmatrix} 0 & 0 & \epsilon \\ 0 & \epsilon^2 & 1 \\ 0 & \epsilon^2 & 1 \end{pmatrix}$	$\begin{pmatrix} 0 & \epsilon & \epsilon \\ 1 & 1 & 1 \\ 0 & \epsilon & \epsilon \end{pmatrix}$	$\begin{pmatrix} 1 & \epsilon & 1 \\ \epsilon & 1 & \epsilon \\ 1 & \epsilon & 1 \end{pmatrix}$	$(\epsilon, 1, \epsilon)$ $(\epsilon, 1, 1)$	$(0, \epsilon, \frac{\pi}{4})$ $(\epsilon, \frac{\pi}{4}, 0)$ $(\frac{\pi}{4}, \epsilon^2, \frac{\pi}{4})$ $(\epsilon, \frac{\pi}{4}, \epsilon^2)$

5.2 Essence of EQLC-Mass Matrices

6	$\begin{pmatrix} 0 & 0 & \xi \\ 0 & \epsilon^2 & 1 \\ 0 & \epsilon^2 & 1 \end{pmatrix}$	$\begin{pmatrix} \epsilon & \epsilon & \epsilon \\ \epsilon & 1 & \epsilon^2 \\ \epsilon & \epsilon & \epsilon \end{pmatrix}$	$\begin{pmatrix} \epsilon & \epsilon & \epsilon \\ \epsilon & 1 & \epsilon^2 \\ \epsilon & \epsilon^2 & 1 \end{pmatrix}$	$(\epsilon, 1, \epsilon)$ $(\epsilon, 1, 1)$	$(\epsilon, \xi, \frac{\pi}{4})$ $(\epsilon, \frac{\pi}{4}, \epsilon)$ $(\epsilon, \epsilon, \xi)$ $(\epsilon, \epsilon, \epsilon)$
7	$\begin{pmatrix} 0 & 0 & \xi \\ 0 & \epsilon^2 & 1 \\ 0 & \epsilon^2 & 1 \end{pmatrix}$	$\begin{pmatrix} \epsilon & \epsilon & \epsilon \\ \epsilon & 1 & \epsilon \\ \epsilon & \epsilon & \epsilon \end{pmatrix}$	$\begin{pmatrix} \epsilon & \epsilon & \epsilon \\ \epsilon & 1 & \epsilon^2 \\ \epsilon & \epsilon^2 & 1 \end{pmatrix}$	$(\epsilon, 1, \epsilon)$ $(\epsilon, 1, 1)$	$(\epsilon, \xi, \frac{\pi}{4})$ $(\epsilon, \frac{\pi}{4}, \epsilon)$ $(\epsilon, \epsilon, \epsilon)$ $(\epsilon, \epsilon, \xi)$
8	$\begin{pmatrix} 0 & 0 & \epsilon \\ 0 & \epsilon^2 & 1 \\ 0 & \epsilon^2 & 1 \end{pmatrix}$	$\begin{pmatrix} \epsilon & \epsilon & \xi \\ \epsilon & \epsilon & \xi \\ \xi & \xi & 1 \end{pmatrix}$	$\begin{pmatrix} \epsilon & \epsilon & 0 \\ \epsilon & 1 & 0 \\ 0 & 0 & 1 \end{pmatrix}$	$(\epsilon, \epsilon, 1)$ $(\epsilon, 1, 1)$	$(\epsilon, \epsilon, \frac{\pi}{4})$ $(\frac{\pi}{4}, \xi, \xi)$ (ϵ, ξ, ξ) $(\epsilon, \epsilon, \frac{\pi}{4})$
9	$\begin{pmatrix} 0 & \epsilon^2 & \epsilon \\ 0 & \epsilon^2 & 1 \\ 0 & \epsilon^2 & 1 \end{pmatrix}$	$\begin{pmatrix} 1 & 1 & 1 \\ \epsilon^2 & \epsilon & \epsilon \\ 1 & 1 & 1 \end{pmatrix}$	$\begin{pmatrix} 1 & \xi & 1 \\ \xi & 1 & \xi \\ 1 & \xi & 1 \end{pmatrix}$	$(1, \epsilon, \epsilon)$ $(\epsilon, 1, 1)$	$(\frac{\pi}{4}, \epsilon, \frac{\pi}{4})$ $(\xi, \frac{\pi}{4}, \xi)$ $(\epsilon, \frac{\pi}{4}, \frac{\pi}{4})$ $(\xi, \frac{\pi}{4}, \xi)$
10	$\begin{pmatrix} 0 & 0 & \xi \\ 0 & \epsilon^2 & 1 \\ 0 & \epsilon^2 & 1 \end{pmatrix}$	$\begin{pmatrix} \epsilon & \epsilon & \epsilon \\ \epsilon^2 & 1 & \epsilon \\ \epsilon & \epsilon & \epsilon \end{pmatrix}$	$\begin{pmatrix} \epsilon & 0 & \xi \\ 0 & 1 & 0 \\ \xi & 0 & 1 \end{pmatrix}$	$(\epsilon, 1, \epsilon)$ $(\epsilon, 1, 1)$	$(\epsilon, \xi, \frac{\pi}{4})$ $(\epsilon, \frac{\pi}{4}, \xi)$ $(0, 0, \epsilon)$ $(0, \xi, \xi)$
11	$\begin{pmatrix} 0 & 0 & \xi \\ 0 & \epsilon^2 & 1 \\ 0 & \epsilon^2 & 1 \end{pmatrix}$	$\begin{pmatrix} \epsilon & \epsilon & \epsilon \\ \epsilon^2 & 1 & \xi \\ \epsilon & \epsilon & \epsilon \end{pmatrix}$	$\begin{pmatrix} \epsilon & 0 & \xi \\ 0 & 1 & 0 \\ \xi & 0 & 1 \end{pmatrix}$	$(\epsilon, 1, \epsilon)$ $(\epsilon, 1, 1)$	$(\epsilon, \xi, \frac{\pi}{4})$ $(\epsilon, \frac{\pi}{4}, \xi)$ $(0, 0, \xi)$ $(0, \xi, \epsilon)$
12	$\begin{pmatrix} 0 & \epsilon^2 & 1 \\ 0 & \epsilon^2 & 1 \\ 0 & \epsilon^2 & 1 \end{pmatrix}$	$\begin{pmatrix} \epsilon & 0 & \xi \\ 0 & \epsilon & \epsilon \\ \xi & \epsilon & 1 \end{pmatrix}$	$\begin{pmatrix} \epsilon & \xi & \xi \\ \xi & 1 & 0 \\ \xi & 0 & 1 \end{pmatrix}$	$(\epsilon, \epsilon, 1)$ $(\epsilon, 1, 1)$	$(\frac{\pi}{4}, \frac{\pi}{4}, \frac{\pi}{4})$ (ξ, ξ, ϵ) (ξ, ξ, ϵ) (ξ, ξ, ϵ)
13	$\begin{pmatrix} 0 & 0 & \epsilon \\ 0 & \epsilon^2 & 1 \\ 0 & \epsilon^2 & 1 \end{pmatrix}$	$\begin{pmatrix} \epsilon & \epsilon & \epsilon \\ \epsilon^2 - \xi & 1 & \xi \\ \epsilon & \epsilon & \epsilon \end{pmatrix}$	$\begin{pmatrix} \epsilon & 0 & \epsilon \\ 0 & 1 & 0 \\ \epsilon & 0 & 1 \end{pmatrix}$	$(\epsilon, 1, \epsilon)$ $(\epsilon, 1, 1)$	$(\xi, \epsilon, \frac{\pi}{4})$ $(\epsilon, \frac{\pi}{4}, \xi)$ (ξ, ξ, ξ) $(\epsilon, \epsilon, \frac{\pi}{4})$
14	$\begin{pmatrix} 0 & \epsilon^2 & 1 \\ 0 & \epsilon^2 & 1 \\ 0 & \epsilon^2 & 1 \end{pmatrix}$	$\begin{pmatrix} \epsilon & \epsilon^2 & \epsilon^2 \\ \epsilon & 1 & \epsilon^2 - \xi \\ \epsilon^2 & \epsilon & \epsilon \end{pmatrix}$	$\begin{pmatrix} \epsilon & \epsilon & \epsilon \\ \epsilon & 1 & \epsilon^2 \\ \epsilon & \epsilon^2 & 1 \end{pmatrix}$	$(\epsilon, 1, \epsilon)$ $(\epsilon, 1, 1)$	$(\frac{\pi}{4}, \frac{\pi}{4}, \frac{\pi}{4})$ $(\xi, \epsilon, \epsilon)$ (ϵ, ξ, ξ) $(\epsilon, \epsilon, \xi)$

15	$\begin{pmatrix} 0 & \epsilon^2 & 1 \\ 0 & \epsilon^2 & 1 \\ 0 & \epsilon^2 & 1 \end{pmatrix}$	$\begin{pmatrix} \epsilon & 0 & \epsilon \\ \epsilon^2 & 1 & \epsilon^2 \\ \epsilon & \epsilon & \epsilon \end{pmatrix}$	$\begin{pmatrix} 1 & \epsilon & 1 \\ \epsilon & 1 & \epsilon \\ 1 & \epsilon & 1 \end{pmatrix}$	$(\epsilon, 1, \epsilon)$ $(\epsilon, 1, 1)$	$(\frac{\pi}{4}, \frac{\pi}{4}, \frac{\pi}{4})$ $(0, \frac{\pi}{4}, \epsilon)$ $(\epsilon^2, \epsilon^2, \epsilon^2)$ $(\epsilon, \frac{\pi}{4}, \epsilon)$
16	$\begin{pmatrix} 0 & \epsilon^2 & 1 \\ 0 & \epsilon^2 & 1 \\ 0 & \epsilon^2 & 1 \end{pmatrix}$	$\begin{pmatrix} \epsilon & \epsilon & \epsilon^2 \\ 1 & 1 & \epsilon^2 \\ \epsilon & \epsilon & \epsilon \end{pmatrix}$	$\begin{pmatrix} 1 & 1 & \epsilon \\ 1 & 1 & \epsilon \\ \epsilon & \epsilon & 1 \end{pmatrix}$	$(\epsilon, 1, \epsilon)$ $(\epsilon, 1, 1)$	$(\frac{\pi}{4}, \frac{\pi}{4}, \frac{\pi}{4})$ $(\epsilon^2, \epsilon, \epsilon)$ $(\frac{\pi}{4}, \epsilon^2, 0)$ $(\frac{\pi}{4}, \epsilon, \epsilon^2)$
17	$\begin{pmatrix} 0 & \epsilon^2 & 1 \\ 0 & \epsilon^2 & \epsilon \\ 0 & \epsilon^2 & 1 \end{pmatrix}$	$\begin{pmatrix} \epsilon & \epsilon^2 & \epsilon \\ 1 & \epsilon & 1 \\ 1 & \epsilon & 1 \end{pmatrix}$	$\begin{pmatrix} 1 & \epsilon & 1 \\ \epsilon & 1 & \epsilon \\ 1 & \epsilon & 1 \end{pmatrix}$	$(\epsilon, \epsilon, 1)$ $(\epsilon, 1, 1)$	$(\frac{\pi}{4}, \frac{\pi}{4}, \epsilon)$ $(\epsilon, \epsilon^2, \frac{\pi}{4})$ $(0, \frac{\pi}{4}, \epsilon^2)$ $(\epsilon, \frac{\pi}{4}, \epsilon^2)$
18	$\begin{pmatrix} 0 & \epsilon^2 & 1 \\ 0 & \epsilon^2 & 1 \\ 0 & \epsilon^2 & 1 \end{pmatrix}$	$\begin{pmatrix} \epsilon & \epsilon & 0 \\ \epsilon & \epsilon & \epsilon \\ \epsilon^2 & \epsilon^2 & 1 \end{pmatrix}$	$\begin{pmatrix} 1 & 1 & 0 \\ 1 & 1 & 0 \\ 0 & 0 & 1 \end{pmatrix}$	$(\epsilon, \epsilon, 1)$ $(\epsilon, 1, 1)$	$(\frac{\pi}{4}, \frac{\pi}{4}, \frac{\pi}{4})$ $(\frac{\pi}{4}, 0, \epsilon)$ $(\epsilon^2, \epsilon^2, \epsilon^2)$ $(\frac{\pi}{4}, \epsilon, \epsilon)$
19	$\begin{pmatrix} 0 & \epsilon^2 & 1 \\ 0 & \epsilon^2 & 1 \\ 0 & \epsilon^2 & 1 \end{pmatrix}$	$\begin{pmatrix} \epsilon & \epsilon^2 & \epsilon \\ \epsilon & \epsilon & \epsilon \\ 1 & \epsilon & 1 \end{pmatrix}$	$\begin{pmatrix} 1 & 0 & 1 \\ 0 & 1 & 0 \\ 1 & 0 & 1 \end{pmatrix}$	$(\epsilon, \epsilon, 1)$ $(\epsilon, 1, 1)$	$(\frac{\pi}{4}, \frac{\pi}{4}, \frac{\pi}{4})$ $(\epsilon, \epsilon^2, \epsilon)$ $(0, \frac{\pi}{4}, \epsilon)$ $(0, \frac{\pi}{4}, 0)$
20	$\begin{pmatrix} 0 & 0 & \epsilon \\ 0 & \epsilon^2 & \xi \\ 0 & 0 & 1 \end{pmatrix}$	$\begin{pmatrix} \epsilon & \epsilon & \epsilon \\ \epsilon & 1 & 1 \\ \epsilon & 1 & 1 \end{pmatrix}$	$\begin{pmatrix} \epsilon & \epsilon & \epsilon \\ \epsilon & 1 & \epsilon^2 \\ \epsilon & \epsilon^2 & 1 \end{pmatrix}$	$(\epsilon, 1, \epsilon)$ $(\epsilon, 1, 1)$	(ξ, ϵ, ξ) $(\epsilon, \epsilon, \frac{\pi}{4})$ $(0, \frac{\pi}{4}, \frac{\pi}{4})$ $(\epsilon, \epsilon, \epsilon)$
21	$\begin{pmatrix} 0 & \epsilon^2 & 1 \\ 0 & \epsilon^2 & \epsilon \\ 0 & \epsilon^2 & 1 \end{pmatrix}$	$\begin{pmatrix} \epsilon & \epsilon^2 & \epsilon^2 \\ \epsilon & 1 & 1 \\ \epsilon & 1 & 1 \end{pmatrix}$	$\begin{pmatrix} \epsilon & \epsilon & \epsilon \\ \epsilon & 1 & \epsilon^2 \\ \epsilon & \epsilon^2 & 1 \end{pmatrix}$	$(\epsilon, 1, \epsilon)$ $(\epsilon, 1, 1)$	$(\frac{\pi}{4}, \frac{\pi}{4}, \epsilon)$ $(0, \frac{\pi}{4}, \frac{\pi}{4})$ $(\epsilon, \frac{\pi}{4}, \frac{\pi}{4})$ $(\epsilon^2, \epsilon, \frac{\pi}{4})$
22	$\begin{pmatrix} 0 & \epsilon^2 & 1 \\ 0 & \epsilon^2 & \epsilon \\ 0 & \epsilon^2 & 1 \end{pmatrix}$	$\begin{pmatrix} \epsilon & 0 & 0 \\ \xi & 1 & \epsilon \\ \xi & 1 & 0 \end{pmatrix}$	$\begin{pmatrix} \epsilon & 0 & \xi \\ 0 & 1 & 0 \\ \xi & 0 & 1 \end{pmatrix}$	$(\epsilon, 1, \epsilon)$ $(\epsilon, 1, 1)$	$(\frac{\pi}{4}, \frac{\pi}{4}, \epsilon)$ $(0, \epsilon, \frac{\pi}{4})$ $(\xi, \epsilon, \epsilon)$ $(\xi, \xi, \frac{\pi}{4})$
23	$\begin{pmatrix} 0 & \epsilon^2 & 1 \\ 0 & \epsilon^2 & \epsilon \\ 0 & \epsilon^2 & 1 \end{pmatrix}$	$\begin{pmatrix} \epsilon & 0 & 0 \\ \xi & 1 & 1 \\ \xi & 1 & 1 \end{pmatrix}$	$\begin{pmatrix} \epsilon & \xi & \xi \\ \xi & 1 & 0 \\ \xi & 0 & 1 \end{pmatrix}$	$(\epsilon, 1, \epsilon)$ $(\epsilon, 1, 1)$	$(\frac{\pi}{4}, \frac{\pi}{4}, \epsilon)$ $(0, \epsilon, \frac{\pi}{4})$ $(\xi, \epsilon, \frac{\pi}{4})$ (ξ, ξ, ϵ)

5.2 Essence of EQLC-Mass Matrices

24	$\begin{pmatrix} 0 & \epsilon^2 & \epsilon \\ 0 & \epsilon^2 & \xi \\ 0 & 0 & 1 \end{pmatrix}$	$\begin{pmatrix} 1 & 1 & 1 \\ 1 & 1 & 1 \\ 1 & 1 & 1 \end{pmatrix}$	$\begin{pmatrix} 1 & 1 & \xi \\ 1 & 1 & \xi \\ \xi & \xi & 1 \end{pmatrix}$	$(1,\epsilon,\epsilon)$ $(\epsilon,1,1)$	$(\frac{\pi}{4},\epsilon,\xi)$ $(\xi,\frac{\pi}{4},\frac{\pi}{4})$ $(\epsilon,\frac{\pi}{4},\frac{\pi}{4})$ $(\frac{\pi}{4},\xi,\xi)$
25	$\begin{pmatrix} 0 & \epsilon^2 & 1 \\ 0 & \epsilon^2 & 1 \\ 0 & \epsilon^2 & 1 \end{pmatrix}$	$\begin{pmatrix} \epsilon & \epsilon^2 & 0 \\ \epsilon & 1 & \epsilon \\ \epsilon^2 & \epsilon & \epsilon \end{pmatrix}$	$\begin{pmatrix} \epsilon & \epsilon & \epsilon^2 \\ \epsilon & 1 & 0 \\ \epsilon^2 & 0 & 1 \end{pmatrix}$	$(\epsilon,1,\epsilon)$ $(\epsilon,1,1)$	$(\frac{\pi}{4},\frac{\pi}{4},\frac{\pi}{4})$ $(0,\frac{\pi}{4},\epsilon)$ $(\epsilon,\frac{\pi}{4},\epsilon)$ $(\epsilon,0,\epsilon)$
26	$\begin{pmatrix} 0 & \epsilon^2 & 1 \\ 0 & \epsilon^2 & \epsilon \\ 0 & \epsilon^2 & 1 \end{pmatrix}$	$\begin{pmatrix} \epsilon & \epsilon^2 & 0 \\ \epsilon & 1 & \epsilon \\ \epsilon & 1 & \epsilon \end{pmatrix}$	$\begin{pmatrix} \epsilon & \epsilon & \epsilon^2-\xi \\ \epsilon & 1 & 0 \\ \epsilon^2-\xi & 0 & 1 \end{pmatrix}$	$(\epsilon,1,\epsilon)$ $(\epsilon,1,1)$	$(\frac{\pi}{4},\frac{\pi}{4},\epsilon)$ $(0,\frac{\pi}{4},\frac{\pi}{4})$ $(\epsilon,\frac{\pi}{4},\epsilon)$ (ϵ,ξ,ϵ)
27	$\begin{pmatrix} 0 & 0 & \epsilon \\ 0 & \epsilon^2 & \xi \\ 0 & 0 & 1 \end{pmatrix}$	$\begin{pmatrix} \epsilon & \epsilon & \xi \\ \epsilon & \epsilon & 1 \\ \epsilon & \epsilon & 1 \end{pmatrix}$	$\begin{pmatrix} \epsilon & \epsilon & 0 \\ \epsilon & 1 & 0 \\ 0 & 0 & 1 \end{pmatrix}$	$(\epsilon,\epsilon,1)$ $(\epsilon,1,1)$	(ϵ,ϵ,ξ) $(\frac{\pi}{4},\xi,\frac{\pi}{4})$ (ϵ,ξ,ξ) $(\epsilon,\epsilon,\frac{\pi}{4})$
28	$\begin{pmatrix} 0 & \epsilon^2 & 1 \\ 0 & \epsilon^2 & 1 \\ 0 & \epsilon^2 & 1 \end{pmatrix}$	$\begin{pmatrix} \epsilon & \epsilon^2 & \epsilon^2 \\ \epsilon & 1 & 1 \\ \epsilon^2 & 0 & \epsilon \end{pmatrix}$	$\begin{pmatrix} \epsilon & \epsilon^2-\xi & \epsilon \\ \epsilon^2-\xi & 1 & 0 \\ \epsilon & 0 & 1 \end{pmatrix}$	$(\epsilon,1,\epsilon)$ $(\epsilon,1,1)$	$(\frac{\pi}{4},\frac{\pi}{4},\frac{\pi}{4})$ $(0,\xi,\epsilon)$ $(\epsilon,\xi,\frac{\pi}{4})$ (ξ,ϵ,ϵ)
29	$\begin{pmatrix} 0 & 0 & 1 \\ 0 & \epsilon^2 & \epsilon \\ 0 & 0 & 1 \end{pmatrix}$	$\begin{pmatrix} \epsilon & 1 & \epsilon \\ 0 & 1 & \xi \\ \epsilon & 1 & \epsilon \end{pmatrix}$	$\begin{pmatrix} 1 & \xi & 1 \\ \xi & 1 & \xi \\ 1 & \xi & 1 \end{pmatrix}$	$(\epsilon,1,\epsilon)$ $(\epsilon,1,1)$	$(\epsilon,\frac{\pi}{4},\epsilon)$ $(\frac{\pi}{4},\frac{\pi}{4},\xi)$ (ϵ,ϵ,ξ) $(\xi,\frac{\pi}{4},\xi)$
30	$\begin{pmatrix} 0 & 0 & 1 \\ 0 & \epsilon^2 & \xi \\ 0 & 0 & 1 \end{pmatrix}$	$\begin{pmatrix} \epsilon & 1 & \epsilon \\ \epsilon^2 & 1 & \epsilon^2 \\ \epsilon & 1 & \epsilon \end{pmatrix}$	$\begin{pmatrix} 1 & \xi & 1 \\ \xi & 1 & \xi \\ 1 & \xi & 1 \end{pmatrix}$	$(\epsilon,1,\epsilon)$ $(\epsilon,1,1)$	$(\epsilon,\frac{\pi}{4},\xi)$ $(\frac{\pi}{4},\frac{\pi}{4},\epsilon)$ (ϵ,ϵ,ξ) $(\xi,\frac{\pi}{4},\xi)$
31	$\begin{pmatrix} 0 & 0 & 1 \\ 0 & \epsilon^2 & \xi \\ 0 & 0 & 1 \end{pmatrix}$	$\begin{pmatrix} \epsilon & 1 & \epsilon \\ \epsilon & 1 & \epsilon^2 \\ \epsilon & 1 & \epsilon \end{pmatrix}$	$\begin{pmatrix} 1 & \epsilon & 1 \\ \epsilon & 1 & \epsilon \\ 1 & \epsilon & 1 \end{pmatrix}$	$(\epsilon,1,\epsilon)$ $(\epsilon,1,1)$	$(\epsilon,\frac{\pi}{4},\xi)$ $(\frac{\pi}{4},\frac{\pi}{4},\epsilon)$ $(\xi,\epsilon,0)$ $(\xi,\frac{\pi}{4},\epsilon)$
32	$\begin{pmatrix} 0 & \epsilon^2 & 1 \\ 0 & \epsilon^2 & \epsilon \\ 0 & \epsilon^2 & 1 \end{pmatrix}$	$\begin{pmatrix} \epsilon & \epsilon & 0 \\ 1 & 1 & \epsilon \\ 1 & 1 & \epsilon \end{pmatrix}$	$\begin{pmatrix} 1 & 1 & \xi \\ 1 & 1 & \xi \\ \xi & \xi & 1 \end{pmatrix}$	$(\epsilon,1,\epsilon)$ $(\epsilon,1,1)$	$(\frac{\pi}{4},\frac{\pi}{4},\epsilon)$ $(\xi,0,\frac{\pi}{4})$ $(\frac{\pi}{4},\xi,\xi)$ $(\frac{\pi}{4},\xi,\xi)$

33	$\begin{pmatrix} 0 & 0 & \xi \\ 0 & \epsilon^2 & \xi \\ 0 & 0 & 1 \end{pmatrix}$	$\begin{pmatrix} \epsilon & \epsilon & \epsilon \\ \epsilon & 1 & 1 \\ \epsilon & 1 & 1 \end{pmatrix}$	$\begin{pmatrix} \epsilon & \xi & \xi \\ \xi & 1 & 0 \\ \xi & 0 & 1 \end{pmatrix}$	$(\epsilon,1,\epsilon)$ $(\epsilon,1,1)$	(ξ,ξ,ξ) $(\xi,\frac{\pi}{4},\frac{\pi}{4})$ $(0,\epsilon,\frac{\pi}{4})$ (ξ,ξ,ξ)
34	$\begin{pmatrix} 0 & 0 & \xi \\ 0 & \epsilon^2 & \xi \\ 0 & 0 & 1 \end{pmatrix}$	$\begin{pmatrix} \epsilon & \xi & \epsilon \\ \epsilon & 1 & \epsilon \\ \epsilon & 1 & \epsilon \end{pmatrix}$	$\begin{pmatrix} \epsilon & \xi & 0 \\ \xi & 1 & 0 \\ 0 & 0 & 1 \end{pmatrix}$	$(\epsilon,1,\epsilon)$ $(\epsilon,1,1)$	(ξ,ξ,ξ) $(\xi,\frac{\pi}{4},\frac{\pi}{4})$ $(0,\epsilon,\xi)$ $(\xi,\xi,\frac{\pi}{4})$
35	$\begin{pmatrix} 0 & 0 & \epsilon \\ 0 & \epsilon^2 & \xi \\ 0 & 0 & 1 \end{pmatrix}$	$\begin{pmatrix} \epsilon & \epsilon & \epsilon \\ 1 & 1 & 1 \\ 1 & 1 & 1 \end{pmatrix}$	$\begin{pmatrix} 1 & \epsilon & 1 \\ \epsilon & 1 & \epsilon \\ 1 & \epsilon & 1 \end{pmatrix}$	$(\epsilon,1,\epsilon)$ $(\epsilon,1,1)$	$(0,\epsilon,\xi)$ $(\epsilon,\frac{\pi}{4},\frac{\pi}{4})$ $(\frac{\pi}{4},\frac{\pi}{4},0)$ $(\epsilon,\frac{\pi}{4},\epsilon)$
36	$\begin{pmatrix} 0 & 0 & 1 \\ 0 & \epsilon^2 & \xi \\ 0 & 0 & 1 \end{pmatrix}$	$\begin{pmatrix} \epsilon & 1 & \epsilon \\ 0 & 1 & \epsilon \\ \epsilon & 1 & \epsilon \end{pmatrix}$	$\begin{pmatrix} \epsilon & \epsilon & \epsilon \\ \epsilon & 1 & \epsilon^2 \\ \epsilon & \epsilon^2 & 1 \end{pmatrix}$	$(\epsilon,1,\epsilon)$ $(\epsilon,1,1)$	$(0,\frac{\pi}{4},\xi)$ $(\frac{\pi}{4},\frac{\pi}{4},\xi)$ $(\epsilon,\frac{\pi}{4},\epsilon)$ $(\epsilon,\epsilon,0)$
37	$\begin{pmatrix} 0 & 0 & 1 \\ 0 & \epsilon^2 & \xi \\ 0 & 0 & 1 \end{pmatrix}$	$\begin{pmatrix} \epsilon & 1 & \epsilon \\ 0 & 1 & 0 \\ \epsilon & 1 & \epsilon \end{pmatrix}$	$\begin{pmatrix} \epsilon & \epsilon & \epsilon \\ \epsilon & 1 & \epsilon^2 \\ \epsilon & \epsilon^2 & 1 \end{pmatrix}$	$(\epsilon,1,\epsilon)$ $(\epsilon,1,1)$	$(0,\frac{\pi}{4},\xi)$ $(\frac{\pi}{4},\frac{\pi}{4},\xi)$ $(\epsilon,\frac{\pi}{4},0)$ $(\epsilon,\epsilon,\epsilon)$
38	$\begin{pmatrix} 0 & \epsilon^2 & 1 \\ 0 & \epsilon^2 & \epsilon \\ 0 & \epsilon^2 & 1 \end{pmatrix}$	$\begin{pmatrix} \epsilon & 0 & \epsilon^2 \\ \epsilon^2-\xi & 1 & \epsilon \\ \epsilon^2 & 1 & \epsilon \end{pmatrix}$	$\begin{pmatrix} \epsilon & \epsilon^2 & \epsilon \\ \epsilon^2 & 1 & 0 \\ \epsilon & 0 & 1 \end{pmatrix}$	$(\epsilon,1,\epsilon)$ $(\epsilon,1,1)$	$(\frac{\pi}{4},\frac{\pi}{4},\epsilon)$ $(0,\epsilon,\frac{\pi}{4})$ (ξ,ξ,ξ) (ξ,ϵ,ϵ)
39	$\begin{pmatrix} 0 & \epsilon^2 & 1 \\ 0 & \epsilon^2 & \epsilon \\ 0 & \epsilon^2 & 1 \end{pmatrix}$	$\begin{pmatrix} \epsilon & 0 & \epsilon^2 \\ \epsilon^2-\xi & 1 & 0 \\ \epsilon^2 & 1 & \epsilon \end{pmatrix}$	$\begin{pmatrix} \epsilon & \xi & \epsilon \\ \xi & 1 & 0 \\ \epsilon & 0 & 1 \end{pmatrix}$	$(\epsilon,1,\epsilon)$ $(\epsilon,1,1)$	$(\frac{\pi}{4},\frac{\pi}{4},\epsilon)$ $(0,\epsilon,\frac{\pi}{4})$ (ξ,ξ,ϵ) (ξ,ϵ,ξ)
40	$\begin{pmatrix} 0 & \epsilon^2 & 1 \\ 0 & \epsilon^2 & \epsilon \\ 0 & \epsilon^2 & 1 \end{pmatrix}$	$\begin{pmatrix} \epsilon & \epsilon^2 & 0 \\ \epsilon^2-\xi & 0 & 1 \\ \epsilon^2 & \epsilon & 1 \end{pmatrix}$	$\begin{pmatrix} \epsilon & \epsilon & \xi \\ \epsilon & 1 & 0 \\ \xi & 0 & 1 \end{pmatrix}$	$(\epsilon,\epsilon,1)$ $(\epsilon,1,1)$	$(\frac{\pi}{4},\frac{\pi}{4},\epsilon)$ $(\epsilon,0,\frac{\pi}{4})$ (ξ,ξ,ϵ) (ϵ,ξ,ξ)
41	$\begin{pmatrix} 0 & 0 & 1 \\ 0 & \epsilon^2 & \xi \\ 0 & 0 & 1 \end{pmatrix}$	$\begin{pmatrix} \epsilon & 1 & 1 \\ \epsilon & 1 & 1 \\ \epsilon & 1 & 1 \end{pmatrix}$	$\begin{pmatrix} \epsilon & \epsilon^2 & \epsilon \\ \epsilon^2 & 1 & 0 \\ \epsilon & 0 & 1 \end{pmatrix}$	$(\epsilon,1,\epsilon)$ $(\epsilon,1,1)$	$(0,\frac{\pi}{4},\xi)$ $(\frac{\pi}{4},\frac{\pi}{4},\xi)$ $(\xi,\frac{\pi}{4},\frac{\pi}{4})$ $(0,\epsilon,\epsilon)$

5.2 Essence of EQLC-Mass Matrices

42	$\begin{pmatrix} 0 & 0 & 1 \\ 0 & \epsilon^2 & \xi \\ 0 & 0 & 1 \end{pmatrix}$	$\begin{pmatrix} \epsilon & 1 & 1 \\ \epsilon & 1 & 1 \\ \epsilon & 1 & 1 \end{pmatrix}$	$\begin{pmatrix} \epsilon & \epsilon & \epsilon^2 \\ \epsilon & 1 & 0 \\ \epsilon^2 & 0 & 1 \end{pmatrix}$	$(\epsilon, 1, \epsilon)$ $(\epsilon, 1, 1)$	$(0, \frac{\pi}{4}, \xi)$ $(\frac{\pi}{4}, \frac{\pi}{4}, \xi)$ $(\xi, \frac{\pi}{4}, \frac{\pi}{4})$ $(\epsilon, 0, \epsilon)$
43	$\begin{pmatrix} 0 & 0 & \epsilon \\ 0 & \epsilon^2 & 1 \\ 0 & \epsilon^2 & 1 \end{pmatrix}$	$\begin{pmatrix} \epsilon^2 & \epsilon^2 & \epsilon^2-\xi \\ \epsilon^2-\xi & \epsilon & \epsilon^2 \\ 1 & \xi & 1 \end{pmatrix}$	$\begin{pmatrix} 1 & \xi & 1 \\ \xi & \epsilon & \xi \\ 1 & \xi & 1 \end{pmatrix}$	$(\epsilon^2, \epsilon, 1)$ $(\epsilon^2, \epsilon, 1)$	$(\epsilon, \epsilon, \frac{\pi}{4})$ (ϵ, ξ, ξ) $(\epsilon, \frac{\pi}{4}, \xi)$ $(\xi, \frac{\pi}{4}, \xi)$
44	$\begin{pmatrix} 0 & 0 & \epsilon \\ 0 & \epsilon^2 & \xi \\ 0 & 0 & 1 \end{pmatrix}$	$\begin{pmatrix} \epsilon^2 & \epsilon^2 & \epsilon^2-\xi \\ 1 & \epsilon & 1 \\ 1 & \epsilon & 1 \end{pmatrix}$	$\begin{pmatrix} 1 & \xi & 1 \\ \xi & \epsilon & \xi \\ 1 & \xi & 1 \end{pmatrix}$	$(\epsilon^2, \epsilon, 1)$ $(\epsilon^2, \epsilon, 1)$	$(\epsilon, \epsilon, \xi)$ $(\epsilon, \xi, \frac{\pi}{4})$ $(\epsilon, \frac{\pi}{4}, \xi)$ $(\xi, \frac{\pi}{4}, \xi)$
45	$\begin{pmatrix} 0 & 0 & 1 \\ 0 & \epsilon^2 & \epsilon \\ 0 & 0 & 1 \end{pmatrix}$	$\begin{pmatrix} \epsilon & 1 & 1 \\ \epsilon & 1 & 1 \\ \epsilon & 1 & 1 \end{pmatrix}$	$\begin{pmatrix} \epsilon & \epsilon & \epsilon \\ \epsilon & 1 & \epsilon^2 \\ \epsilon & \epsilon^2 & 1 \end{pmatrix}$	$(\epsilon, 1, \epsilon)$ $(\epsilon, 1, 1)$	$(\epsilon, \frac{\pi}{4}, \epsilon)$ $(\frac{\pi}{4}, \frac{\pi}{4}, \epsilon^2)$ $(\epsilon^2, \frac{\pi}{4}, \frac{\pi}{4})$ $(\epsilon^2, \epsilon, \frac{\pi}{4})$
46	$\begin{pmatrix} 0 & 0 & 0 \\ 0 & \epsilon^2 & \xi \\ 0 & 0 & 1 \end{pmatrix}$	$\begin{pmatrix} \epsilon & 0 & \epsilon \\ 1 & 1 & \epsilon^2 \\ 1 & 1 & \epsilon \end{pmatrix}$	$\begin{pmatrix} 1 & 1 & \xi \\ 1 & 1 & \xi \\ \xi & \xi & 1 \end{pmatrix}$	$(\epsilon, 1, \epsilon)$ $(\epsilon, 1, 1)$	$(\epsilon, 0, \xi)$ $(\epsilon, \frac{\pi}{4}, \frac{\pi}{4})$ $(\frac{\pi}{4}, \epsilon, \xi)$ $(\frac{\pi}{4}, \xi, \xi)$
47	$\begin{pmatrix} 0 & 0 & 0 \\ 0 & \epsilon^2 & \xi \\ 0 & 0 & 1 \end{pmatrix}$	$\begin{pmatrix} \epsilon & 0 & \epsilon \\ 1 & 1 & \epsilon \\ 1 & 1 & \epsilon \end{pmatrix}$	$\begin{pmatrix} 1 & 1 & \epsilon \\ 1 & 1 & \epsilon \\ \epsilon & \epsilon & 1 \end{pmatrix}$	$(\epsilon, 1, \epsilon)$ $(\epsilon, 1, 1)$	$(\epsilon, 0, \xi)$ $(\epsilon, \frac{\pi}{4}, \frac{\pi}{4})$ $(\frac{\pi}{4}, \xi, \xi)$ $(\frac{\pi}{4}, \xi, \epsilon)$
48	$\begin{pmatrix} 0 & 0 & \xi \\ 0 & \epsilon^2 & 1 \\ 0 & \epsilon^2 & 1 \end{pmatrix}$	$\begin{pmatrix} \epsilon^2 & \epsilon & \epsilon \\ \epsilon^2 & \epsilon & \epsilon \\ 0 & \epsilon^2 & 1 \end{pmatrix}$	$\begin{pmatrix} \epsilon^2 & \epsilon^2 & \xi \\ \epsilon^2 & \epsilon & 0 \\ \xi & 0 & 1 \end{pmatrix}$	$(\epsilon^2, \epsilon, 1)$ $(\epsilon^2, \epsilon, 1)$	$(\epsilon, \xi, \frac{\pi}{4})$ $(\frac{\pi}{4}, \epsilon, \epsilon)$ $(\xi, 0, \xi)$ $(\epsilon, \xi, 0)$
49	$\begin{pmatrix} 0 & 0 & \epsilon \\ 0 & \epsilon^2 & 0 \\ 0 & 0 & 1 \end{pmatrix}$	$\begin{pmatrix} \epsilon^2 & \epsilon & 0 \\ \epsilon & 1 & 1 \\ \epsilon & 1 & 1 \end{pmatrix}$	$\begin{pmatrix} \epsilon^2 & \epsilon & \epsilon \\ \epsilon & 1 & 1 \\ \epsilon & 1 & 1 \end{pmatrix}$	$(\epsilon^2, 1, \epsilon)$ $(\epsilon^2, \epsilon, 1)$	$(\epsilon^2, \epsilon, 0)$ $(\epsilon, \frac{\pi}{4}, \frac{\pi}{4})$ $(\epsilon, \epsilon^2, \frac{\pi}{4})$ $(\epsilon^2, \epsilon, \frac{\pi}{4})$
50	$\begin{pmatrix} 0 & 0 & \epsilon \\ 0 & \epsilon^2 & 0 \\ 0 & 0 & 1 \end{pmatrix}$	$\begin{pmatrix} \epsilon^2 & 0 & \epsilon \\ \epsilon & 1 & 1 \\ \epsilon & 1 & 1 \end{pmatrix}$	$\begin{pmatrix} \epsilon^2 & \epsilon & \epsilon \\ \epsilon & 1 & 1 \\ \epsilon & 1 & 1 \end{pmatrix}$	$(\epsilon^2, 1, \epsilon)$ $(\epsilon^2, \epsilon, 1)$	$(\epsilon^2, \epsilon, 0)$ $(\epsilon, \frac{\pi}{4}, \frac{\pi}{4})$ $(\epsilon, \epsilon^2, \frac{\pi}{4})$ $(\epsilon^2, \epsilon, \frac{\pi}{4})$

51	$\begin{pmatrix} 0 & 0 & \xi \\ 0 & \epsilon^2 & \epsilon \\ 0 & 0 & 1 \end{pmatrix}$	$\begin{pmatrix} \epsilon & \epsilon^2 - \xi & \epsilon \\ \epsilon & 1 & \epsilon \\ \epsilon & 1 & \epsilon \end{pmatrix}$	$\begin{pmatrix} \epsilon & \epsilon & \epsilon \\ \epsilon & 1 & \epsilon^2 \\ \epsilon & \epsilon^2 & 1 \end{pmatrix}$	$(\epsilon, 1, \epsilon)$ $(\epsilon, 1, 1)$	(ξ, ξ, ϵ) $(\xi, \frac{\pi}{4}, \frac{\pi}{4})$ (ϵ, ξ, ξ) $(\epsilon, 0, \frac{\pi}{4})$
52	$\begin{pmatrix} 0 & 0 & \xi \\ 0 & \epsilon^2 & \epsilon \\ 0 & 0 & 1 \end{pmatrix}$	$\begin{pmatrix} \epsilon & \epsilon & \epsilon \\ \epsilon & 1 & 1 \\ \epsilon & 1 & 1 \end{pmatrix}$	$\begin{pmatrix} \epsilon & \epsilon & 0 \\ \epsilon & 1 & 0 \\ 0 & 0 & 1 \end{pmatrix}$	$(\epsilon, 1, \epsilon)$ $(\epsilon, 1, 1)$	(ξ, ξ, ϵ) $(\xi, \frac{\pi}{4}, \frac{\pi}{4})$ $(\epsilon, \xi, \frac{\pi}{4})$ $(\epsilon, 0, \xi)$
53	$\begin{pmatrix} 0 & 0 & \xi \\ 0 & \epsilon^2 & \epsilon \\ 0 & 0 & 1 \end{pmatrix}$	$\begin{pmatrix} \epsilon & \epsilon & \epsilon \\ \epsilon & 1 & 1 \\ \epsilon & 1 & 1 \end{pmatrix}$	$\begin{pmatrix} \epsilon & 0 & \epsilon \\ 0 & 1 & 0 \\ \epsilon & 0 & 1 \end{pmatrix}$	$(\epsilon, 1, \epsilon)$ $(\epsilon, 1, 1)$	(ξ, ξ, ϵ) $(\xi, \frac{\pi}{4}, \frac{\pi}{4})$ $(\epsilon, \xi, \frac{\pi}{4})$ $(0, \epsilon, \xi)$
54	$\begin{pmatrix} 0 & \epsilon^2 & \epsilon \\ 0 & \epsilon^2 & \epsilon \\ 0 & 0 & 1 \end{pmatrix}$	$\begin{pmatrix} 1 & 1 & 1 \\ 1 & 1 & 1 \\ 1 & 1 & 1 \end{pmatrix}$	$\begin{pmatrix} 1 & 1 & \epsilon \\ 1 & 1 & \epsilon \\ \epsilon & \epsilon & 1 \end{pmatrix}$	$(1, \epsilon, \epsilon)$ $(\epsilon, 1, 1)$	$(\frac{\pi}{4}, \epsilon, \epsilon)$ $(\epsilon, \frac{\pi}{4}, \frac{\pi}{4})$ $(\epsilon^2, \frac{\pi}{4}, \frac{\pi}{4})$ $(\frac{\pi}{4}, \epsilon, 0)$
55	$\begin{pmatrix} 0 & \epsilon^2 & \epsilon \\ 0 & \epsilon^2 & 1 \\ 0 & \epsilon^2 & 1 \end{pmatrix}$	$\begin{pmatrix} 1 & 1 & 1 \\ \epsilon & \epsilon & \epsilon \\ 1 & 1 & 1 \end{pmatrix}$	$\begin{pmatrix} 1 & 1 & \epsilon \\ 1 & 1 & \epsilon \\ \epsilon & \epsilon & 1 \end{pmatrix}$	$(1, \epsilon, \epsilon)$ $(\epsilon, 1, 1)$	$(\frac{\pi}{4}, \epsilon, \frac{\pi}{4})$ $(\epsilon, \frac{\pi}{4}, \epsilon)$ $(\epsilon^2, \frac{\pi}{4}, \frac{\pi}{4})$ $(\frac{\pi}{4}, \epsilon, 0)$
56	$\begin{pmatrix} 0 & 0 & \epsilon \\ 0 & \epsilon^2 & 1 \\ 0 & \epsilon^2 & 1 \end{pmatrix}$	$\begin{pmatrix} \epsilon^2 & \epsilon & \epsilon \\ 1 & 1 & \epsilon \\ \epsilon^2 & \epsilon & \epsilon \end{pmatrix}$	$\begin{pmatrix} 1 & 1 & \epsilon \\ 1 & 1 & \epsilon \\ \epsilon & \epsilon & 1 \end{pmatrix}$	$(\epsilon, 1, \epsilon)$ $(\epsilon, 1, 1)$	$(\epsilon^2, \epsilon, \frac{\pi}{4})$ $(\epsilon, \frac{\pi}{4}, \epsilon^2)$ $(\frac{\pi}{4}, \epsilon, \epsilon^2)$ $(\frac{\pi}{4}, \epsilon, \epsilon^2)$
57	$\begin{pmatrix} 0 & 0 & \epsilon \\ 0 & \epsilon^2 & \xi \\ 0 & 0 & 1 \end{pmatrix}$	$\begin{pmatrix} \epsilon^2 & \epsilon & \epsilon \\ 1 & 1 & \epsilon \\ 1 & 1 & \zeta \end{pmatrix}$	$\begin{pmatrix} 1 & 1 & \epsilon \\ 1 & 1 & \epsilon \\ \epsilon & \epsilon & 1 \end{pmatrix}$	$(\epsilon, 1, \epsilon)$ $(\epsilon, 1, 1)$	(ξ, ϵ, ξ) $(\epsilon, \frac{\pi}{4}, \frac{\pi}{4})$ $(\frac{\pi}{4}, \epsilon, \xi)$ $(\frac{\pi}{4}, \epsilon, \xi)$
58	$\begin{pmatrix} 0 & 0 & \epsilon \\ 0 & \epsilon^2 & \xi \\ 0 & 0 & 1 \end{pmatrix}$	$\begin{pmatrix} \epsilon^2 & \epsilon & \xi \\ \epsilon & \epsilon & 1 \\ \epsilon & \epsilon & 1 \end{pmatrix}$	$\begin{pmatrix} \epsilon^2 & 0 & \epsilon \\ 0 & \epsilon & \xi \\ \epsilon & \xi & 1 \end{pmatrix}$	$(\epsilon^2, \epsilon, 1)$ $(\epsilon^2, \epsilon, 1)$	$(\epsilon, \epsilon, \xi)$ $(\frac{\pi}{4}, \xi, \frac{\pi}{4})$ (ξ, ϵ, ξ) (ξ, ϵ, ξ)
59	$\begin{pmatrix} 0 & 0 & 1 \\ 0 & \epsilon^2 & 1 \\ 0 & \epsilon^2 & 1 \end{pmatrix}$	$\begin{pmatrix} 1 & \epsilon & 1 \\ 1 & \epsilon^2 & 1 \\ 1 & \epsilon & 1 \end{pmatrix}$	$\begin{pmatrix} 1 & \epsilon & 1 \\ \epsilon & 1 & \epsilon \\ 1 & \epsilon & 1 \end{pmatrix}$	$(1, \epsilon, \epsilon)$ $(\epsilon, 1, 1)$	$(\epsilon, \frac{\pi}{4}, \frac{\pi}{4})$ $(\frac{\pi}{4}, \frac{\pi}{4}, \epsilon)$ $(\epsilon, \frac{\pi}{4}, \epsilon^2)$ $(\epsilon, \frac{\pi}{4}, \epsilon^2)$

5.2 Essence of EQLC-Mass Matrices

#	M_1	M_2	M_3		
60	$\begin{pmatrix} 0 & 0 & 0 \\ 0 & \epsilon^2 & 1 \\ 0 & \epsilon^2 & 1 \end{pmatrix}$	$\begin{pmatrix} 0 & \epsilon & \epsilon \\ 1 & 1 & 0 \\ \xi & \epsilon & \epsilon \end{pmatrix}$	$\begin{pmatrix} 1 & 1 & \epsilon \\ 1 & 1 & \epsilon \\ \epsilon & \epsilon & 1 \end{pmatrix}$	$(\epsilon, 1, \epsilon)$ $(\epsilon, 1, 1)$	$(\epsilon, 0, \frac{\pi}{4})$ $(\epsilon, \frac{\pi}{4}, \xi)$ $(\frac{\pi}{4}, \xi, \xi)$ $(\frac{\pi}{4}, \epsilon, \xi)$
61	$\begin{pmatrix} 0 & 0 & \epsilon \\ 0 & \epsilon^2 & 1 \\ 0 & \epsilon^2 & 1 \end{pmatrix}$	$\begin{pmatrix} \epsilon^2 & 0 & \epsilon \\ \xi & 1 & 1 \\ \epsilon^2 & \epsilon & \xi \end{pmatrix}$	$\begin{pmatrix} \epsilon^2 & 0 & 0 \\ 0 & 1 & 1 \\ 0 & 1 & 1 \end{pmatrix}$	$(\epsilon^2, 1, \epsilon)$ $(\epsilon^2, \epsilon, 1)$	$(\xi, \epsilon, \frac{\pi}{4})$ $(\epsilon, \frac{\pi}{4}, \xi)$ $(\xi, \xi, \frac{\pi}{4})$ $(\xi, 0, \frac{\pi}{4})$
62	$\begin{pmatrix} 0 & 0 & \epsilon \\ 0 & \epsilon^2 & \xi \\ 0 & 0 & 1 \end{pmatrix}$	$\begin{pmatrix} \epsilon^2 & \epsilon & 0 \\ \epsilon^2 & 1 & 1 \\ \epsilon^2 & 1 & 1 \end{pmatrix}$	$\begin{pmatrix} \epsilon^2 & 0 & 0 \\ 0 & 1 & 1 \\ 0 & 1 & 1 \end{pmatrix}$	$(\epsilon^2, 1, \epsilon)$ $(\epsilon^2, \epsilon, 1)$	(ξ, ϵ, ξ) $(\epsilon, \frac{\pi}{4}, \frac{\pi}{4})$ $(\xi, \xi, \frac{\pi}{4})$ $(\xi, 0, \frac{\pi}{4})$
63	$\begin{pmatrix} 0 & 0 & \epsilon \\ 0 & \epsilon^2 & \xi \\ 0 & 0 & 1 \end{pmatrix}$	$\begin{pmatrix} \epsilon^2 & 0 & \epsilon \\ \epsilon^2 & 1 & 1 \\ \epsilon^2 & 1 & 1 \end{pmatrix}$	$\begin{pmatrix} \epsilon^2 & 0 & 0 \\ 0 & 1 & 1 \\ 0 & 1 & 1 \end{pmatrix}$	$(\epsilon^2, 1, \epsilon)$ $(\epsilon^2, \epsilon, 1)$	(ξ, ϵ, ξ) $(\epsilon, \frac{\pi}{4}, \frac{\pi}{4})$ $(\xi, \xi, \frac{\pi}{4})$ $(\xi, 0, \frac{\pi}{4})$
64	$\begin{pmatrix} 0 & 0 & \epsilon \\ 0 & \epsilon^2 & \xi \\ 0 & 0 & 1 \end{pmatrix}$	$\begin{pmatrix} \epsilon & \epsilon & 0 \\ \epsilon & 1 & 1 \\ \epsilon & 1 & 1 \end{pmatrix}$	$\begin{pmatrix} \epsilon & 0 & \epsilon \\ 0 & 1 & 0 \\ \epsilon & 0 & 1 \end{pmatrix}$	$(\epsilon, 1, \epsilon)$ $(\epsilon, 1, 1)$	$(0, \epsilon, \xi)$ $(\epsilon, \frac{\pi}{4}, \frac{\pi}{4})$ $(\epsilon, \epsilon, \frac{\pi}{4})$ $(0, \epsilon, \xi)$
65	$\begin{pmatrix} 0 & \epsilon^2 & \epsilon \\ 0 & \epsilon^2 & 1 \\ 0 & \epsilon^2 & 1 \end{pmatrix}$	$\begin{pmatrix} \epsilon^2 & \xi & 1 \\ 0 & \epsilon & \xi \\ \epsilon^2 & \xi & 1 \end{pmatrix}$	$\begin{pmatrix} \epsilon^2 & \epsilon^2 & \xi \\ \epsilon^2 & \epsilon & 0 \\ \xi & 0 & 1 \end{pmatrix}$	$(\epsilon^2, \epsilon, 1)$ $(\epsilon^2, \epsilon, 1)$	$(\frac{\pi}{4}, \epsilon, \frac{\pi}{4})$ $(\xi, \frac{\pi}{4}, \xi)$ $(\xi, 0, \xi)$ $(\epsilon, \xi, 0)$
66	$\begin{pmatrix} 0 & \epsilon^2 & \epsilon \\ 0 & \epsilon^2 & \xi \\ 0 & 0 & 1 \end{pmatrix}$	$\begin{pmatrix} \epsilon^2 & \xi & 1 \\ \epsilon^2 & \epsilon & 1 \\ \epsilon^2 & \epsilon & 1 \end{pmatrix}$	$\begin{pmatrix} \epsilon^2 & \epsilon^2 & \xi \\ \epsilon^2 & \epsilon & 0 \\ \xi & 0 & 1 \end{pmatrix}$	$(\epsilon^2, \epsilon, 1)$ $(\epsilon^2, \epsilon, 1)$	$(\frac{\pi}{4}, \epsilon, \xi)$ $(\xi, \frac{\pi}{4}, \frac{\pi}{4})$ $(\xi, 0, \xi)$ $(\epsilon, \xi, 0)$
67	$\begin{pmatrix} 0 & 0 & \xi \\ 0 & \epsilon^2 & \epsilon \\ 0 & 0 & 1 \end{pmatrix}$	$\begin{pmatrix} \epsilon & \epsilon^2 & \epsilon \\ 1 & 1 & 1 \\ 1 & 1 & 1 \end{pmatrix}$	$\begin{pmatrix} 1 & \xi & 1 \\ \xi & 1 & \xi \\ 1 & \xi & 1 \end{pmatrix}$	$(\epsilon, 1, \epsilon)$ $(\epsilon, 1, 1)$	$(\epsilon, \xi, \epsilon)$ $(\epsilon, \frac{\pi}{4}, \frac{\pi}{4})$ $(\frac{\pi}{4}, \frac{\pi}{4}, \epsilon)$ $(0, \frac{\pi}{4}, \xi)$
68	$\begin{pmatrix} 0 & 0 & \xi \\ 0 & \epsilon^2 & \epsilon \\ 0 & 0 & 1 \end{pmatrix}$	$\begin{pmatrix} \epsilon & 0 & \epsilon \\ 1 & 1 & 1 \\ 1 & 1 & 1 \end{pmatrix}$	$\begin{pmatrix} 1 & \epsilon & 1 \\ \epsilon & 1 & \epsilon \\ 1 & \epsilon & 1 \end{pmatrix}$	$(\epsilon, 1, \epsilon)$ $(\epsilon, 1, 1)$	$(\epsilon, \xi, \epsilon)$ $(\epsilon, \frac{\pi}{4}, \frac{\pi}{4})$ $(\frac{\pi}{4}, \frac{\pi}{4}, \xi)$ $(0, \frac{\pi}{4}, \epsilon)$

#	M_R	M_D	M_ℓ	m_ν	mixing angles
69	$\begin{pmatrix} 0 & 0 & \xi \\ 0 & \epsilon^2 & 1 \\ 0 & \epsilon^2 & 1 \end{pmatrix}$	$\begin{pmatrix} \epsilon & \epsilon^2 & \epsilon \\ 1 & 1 & 1 \\ \epsilon & \epsilon & \epsilon \end{pmatrix}$	$\begin{pmatrix} 1 & \xi & 1 \\ \xi & 1 & \xi \\ 1 & \xi & 1 \end{pmatrix}$	$(\epsilon,1,\epsilon)$ $(\epsilon,1,1)$	$(\epsilon,\xi,\frac{\pi}{4})$ $(\epsilon,\frac{\pi}{4},\epsilon)$ $(\frac{\pi}{4},\frac{\pi}{4},\epsilon)$ $(0,\frac{\pi}{4},\xi)$
70	$\begin{pmatrix} 0 & 0 & \xi \\ 0 & \epsilon^2 & 1 \\ 0 & \epsilon^2 & 1 \end{pmatrix}$	$\begin{pmatrix} \epsilon & 0 & \epsilon \\ 1 & 1 & 1 \\ \epsilon & \epsilon & \epsilon \end{pmatrix}$	$\begin{pmatrix} 1 & \epsilon & 1 \\ \epsilon & 1 & \epsilon \\ 1 & \epsilon & 1 \end{pmatrix}$	$(\epsilon,1,\epsilon)$ $(\epsilon,1,1)$	$(\epsilon,\xi,\frac{\pi}{4})$ $(\epsilon,\frac{\pi}{4},\epsilon)$ $(\frac{\pi}{4},\frac{\pi}{4},\xi)$ $(0,\frac{\pi}{4},\epsilon)$
71	$\begin{pmatrix} 0 & 0 & \xi \\ 0 & \epsilon^2 & 0 \\ 0 & 0 & 1 \end{pmatrix}$	$\begin{pmatrix} \epsilon^2 & \epsilon & \epsilon^2 \\ 1 & \epsilon & 1 \\ 1 & \epsilon & 1 \end{pmatrix}$	$\begin{pmatrix} 1 & \epsilon & 1 \\ \epsilon & \epsilon & \epsilon \\ 1 & \epsilon & 1 \end{pmatrix}$	$(\epsilon^2,\epsilon,1)$ $(\epsilon^2,\epsilon,1)$	$(\xi,\xi,0)$ $(\frac{\pi}{4},\xi,\frac{\pi}{4})$ $(\xi,\frac{\pi}{4},0)$ $(\xi,\frac{\pi}{4},\epsilon)$
72	$\begin{pmatrix} 0 & 0 & \xi \\ 0 & \epsilon^2 & 0 \\ 0 & 0 & 1 \end{pmatrix}$	$\begin{pmatrix} \epsilon^2 & \epsilon & \epsilon^2 \\ 1 & \epsilon & 1 \\ 1 & 0 & 1 \end{pmatrix}$	$\begin{pmatrix} 1 & 0 & 1 \\ 0 & \epsilon & 0 \\ 1 & 0 & 1 \end{pmatrix}$	$(\epsilon^2,\epsilon,1)$ $(\epsilon^2,\epsilon,1)$	$(\xi,\xi,0)$ $(\frac{\pi}{4},\xi,\frac{\pi}{4})$ $(\xi,\frac{\pi}{4},\epsilon)$ $(\xi,\frac{\pi}{4},0)$

Table 5.1: Complete set of selected seesaw textures/realizations, where $\xi \in \{0,\epsilon^2\}$.

As an outcome of our systematic and model independent bottom-up approach we have found new textures, e.g., the texture of M_R for #72 in Table 5.1, which we call a "diamond texture" due to the enclosing of a diamond shape of the four "1" entries in the corners. Note again, all texture entries with powers of $\epsilon \geq 3$ are approximated by 0 since this corresponds to the experimental error of oscillation experiments. Thereby, we can see that we obtain a broad class of realizations: For example, hierarchical and semi-hierarchical mass spectra for M_D and M_R, shown in Fig. 5.2, and large mixings in the charged lepton and/or neutrino sector. Moreover, maximal mixing can occur for Dirac as well as for Majorana neutrinos as illustrated in Fig. 5.3. All distributions are obtained by simply counting the number of realizations with a certain mass spectrum or mixings. Note that for M_D more mass hierarchies than the dominant $(\epsilon,\epsilon,1)$ and $(\epsilon,1,\epsilon)$ are realized. These are denoted as "Others" in Fig. 5.2. The number of maximal mixing angles in U_D, $U_{D'}$, U_R, and U_ℓ, respectively, is shown in Fig. 5.3, where "All small" refers to mixing matrices with mixing angles smaller than ϵ. In Fig. 5.4 we show the distribution of special cases of all allowed seesaw realizations (not texture sets) such as symmetric M_D (i.e.,

5.2 Essence of EQLC-Mass Matrices

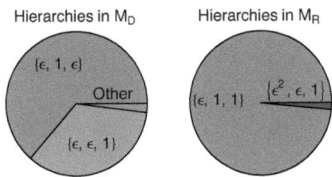

Figure 5.2: Distributions of mass hierarchies in M_D (left) and M_R (right) for all valid seesaw realizations.

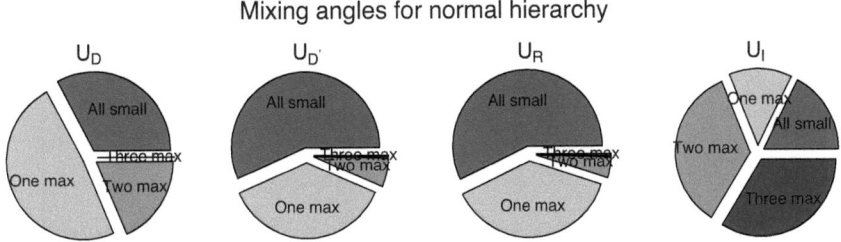

Figure 5.3: Distributions of mixings in U_D, $U_{D'}$, U_R, and U_ℓ, for all valid seesaw realizations. The different pie labels refer to the number of maximal mixing angles, where "All small" corresponds to all mixing angles $\leq \epsilon$.

$U_D \simeq U_{D'}$), diagonal M_R (i.e., $U_R \simeq$), and $U_\ell \simeq$, where "$\simeq$" is valid within our precision of ϵ^2. Thus, we obtain cases often considered in literature such as small mixings in the charged lepton and Majorana sector but also completely new realizations such as three maximal mixings in U_ℓ. Nevertheless, one may conclude from our results that there exist many possibilities to implement the seesaw mechanism without such restrictions on the mass matrices.

The mass spectra of M_R shown in Fig. 5.2 may have immediate relevance for leptogenesis (by taking CP phases into account as we will do in Chap. 7). In more than 80% of the cases, the right-handed neutrino mass spectrum has the semi-hierarchical form $(\epsilon, 1, 1)$. In these cases, successful leptogenesis might be achieved via resonant leptogenesis or by taking flavor effects into account (for a connection with low energy CP violation see, e.g., Ref. [106]). This may lead to testable collider implications if m_3^R is in the resonant limit of TeV range. However, in the literature, usually strongly hierarchical right-handed neutrino masses are considered for leptogenesis. In our analysis, such scenarios are present but found to be by a factor of about 5 less abundant than for mild hierarchy. Nota bene, the strict hierarchical mass spectra of M_R, i.e., $(\epsilon^2, \epsilon, 1)$, can in some cases be amplified to $(\epsilon^n, \epsilon, 1)$, where $n \geq 2$, without the need for other

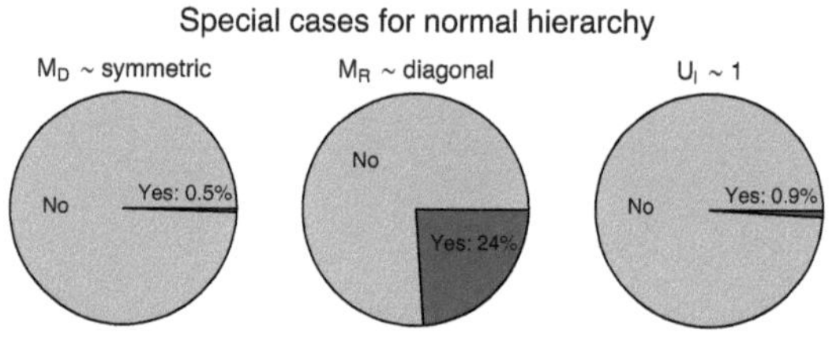

Figure 5.4: Fraction of special cases such as symmetric M_D, diagonal M_R, and $U_\ell \simeq$, of all valid realizations (not texture sets) up to ϵ^2 precision.

modifications. A sufficiently large n, e.g., $n = 8$, can allow a seesaw scale of $m_3^R \sim 10^{14}$ GeV for the strongly hierarchical case and could generate sufficient baryon asymmetry through flavored leptogenesis.

5.2.2 Performance

In this section, we would like to discuss the performance of the seesaw realizations shown in Tables 5.1 and A.1. All have in common that they resist an increased experimental pressure if θ_{13} will not be measured. Consequently, θ_{13} is not a restrictive selection criterion (this may change for mass matrix textures constructed with other PMNS values than given in Eq. (4.3)). For this we introduce a performance indicator χ^2 corresponding to a Gaussian χ^2 approximation in $\sin^2 \theta_{12}$ and $\sin^2 \theta_{23}$ with the current best-fit values, which is defined as

$$\chi^2 \equiv \left(\frac{\sin^2 \theta_{12} - 0.3}{0.3 \times \sigma_{12}} \right)^2 + \left(\frac{\sin^2 \theta_{23} - 0.5}{0.5 \times \sigma_{23}} \right)^2 . \quad (5.4)$$

This reflects the accordance of a realization with current experimental data, e.g., $\chi^2 = 11.83$ corresponds to a 3σ CL exclusion for 2 degrees of freedom (d.o.f.).[5] In Fig. 5.5, we show the distribution of all valid seesaw realizations as a function of χ^2 defined in Eq. (5.4). Obviously, our approach presented in Sec. 5.1.2 already ensures the compatibility of each realization with current experimental bounds. Moreover, in all valid cases $\theta_{13} \ll 1°$ and only 6.5% of the realizations lead to $11.83 \lesssim \chi^2 \lesssim 17$ (corresponding to a CL between 3 and 4σ for 2 d.o.f.). In other words, all presented realizations are in perfect agreement with current data and lead to

---

[5]We use $\sigma_{12} \simeq 9\%$ (for $\sin^2 \theta_{12}$) and $\sigma_{23} \simeq 16\%$ (for $\sin^2 \theta_{23}$) for the relative 1σ errors [51]. Note, we only find $\sin^2 \theta_{13} \ll 0.04$ which is below the current bound, i.e., we do not have to impose an additional selection criterion for θ_{13}.

5.2 Essence of EQLC-Mass Matrices

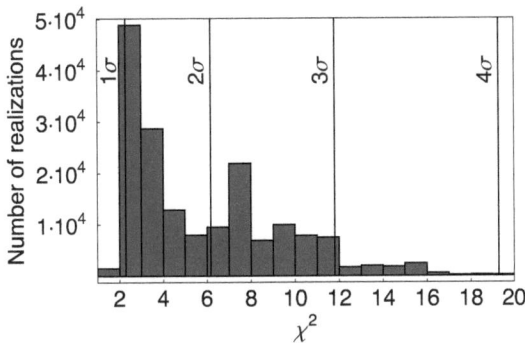

Figure 5.5: Distribution of valid seesaw realizations as a function of χ^2 as defined in Eq. (5.4). The values of θ_{13} are all in agreement with current data.

nearly tribimaximal mixing, *i.e.*, very small θ_{13}. This might naively be expected since for our matching in $M_{\text{eff}}^{\text{exp}}$ we have used the best-fit values. However, we find realizations with small $\theta_{13} \simeq 0°$ but in almost all cases we obtain $\theta_{23} \simeq 50°$ despite of the best-fit input value of $45°$. This deviation from maximal atmospheric mixing can be tested at 3σ CL by the T2K and NOνA experiments [72] and the sign of the deviation from maximal mixing with a neutrino factory at 3σ CL for $\sin^2 2\theta_{13} \gtrsim 10^{-2.5}$, or at the 90% CL otherwise [73].

Above, we have discussed the "quality" of the realizations with respect to experimental data. In the following, we concentrate on the naturalness of the Yukawa couplings. Since our approach to obtain the mass matrix sets did not involve any constraints on the Yukawas they are rather an outcome of our hypothesis of EQLC. Therefore, we check whether they are unnaturally small or large. A proper coupling should for us be of order one, *i.e.*, between ϵ and $1/\epsilon$, since the identification of the leading term of the expansion in ϵ (thus the texture entry) would otherwise not be meaningful. In Fig. 5.6, we show the distribution of the absolute values of the Yukawa couplings for M_ℓ, M_D, and M_R, respectively. For this, we have analytically expanded each mass matrix element in ϵ up to order two and have considered the factors for each order and not only the leading one. In Fig. 5.6, we can see that in 99.9% of all cases the factors are of order one. Therefore, the name "order one factor" ("order one coupling" for the leading term) is justified. Note, although the first bin corresponds to factors smaller than ϵ the texture extraction is unambiguous since in such cases all other factors become also very small and not only the leading one.

Now, we discuss the stability of our results with respect to RG running from high-scale

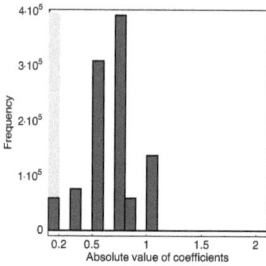 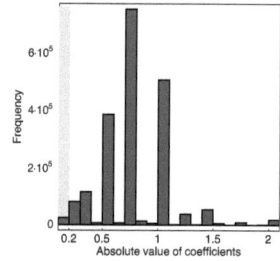 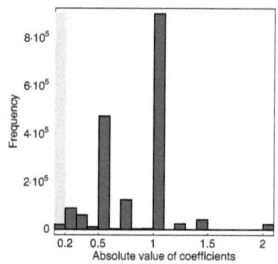

Figure 5.6: Distribution of the Yukawa couplings of M_ℓ (left), M_D (center), and M_R (right) for all valid seesaw realizations (of all orders and for all matrix elements). The gray-shaded region marks the area for couplings smaller than ϵ.

to $\sim 1\,\text{GeV}$. All obtained seesaw realizations are based on our hypothesis of extended QLC holding at high energies and are compared in our method with low energy data. This may cause problems. First of all, we take a look at θ_C. The Cabibbo angle exhibits no running [107] and $V_{cb} \sim \epsilon^2$ typically changes only by a factor smaller than 2 when running from $\sim 1\,\text{GeV}$ up to the Planck scale $\sim 10^{19}\,\text{GeV}$ [107, 108]. Next, note that also the running of a possibly maximal atmospheric mixing angle θ_{23} is negligible (due to the smallness of the charged lepton Yukawa couplings), unless we work in the MSSM with large $\tan\beta$ [72]. For the running of M_{eff} from the GUT scale down to low energies, the corrections to the leptonic mixing angle θ_{ij} are smaller than $\sim |m_i + m_j|^2(|m_i|^2 - |m_j|^2)^{-1} \times 10^{-2}$, where m_i and m_j are the eigenvalues of the ith and jth neutrino mass eigenstates of M_{eff} at the GUT scale [109]. For NH neutrinos, the corrections are thus $\lesssim 1°$, i.e., negligible. This may change for IH or the degenerate case. However, note that independent of the neutrino mass hierarchies, it is always possible to switch off RG effects on neutrino mixing angles by tuning the phases [109]. In addition, while the absolute neutrino mass scale is affected by RG running, the neutrino mass ratios are hardly changed. Summarizing, since our results are very stable under RG running for NH neutrinos we (rightly) neglect it (for a more detailed discussion and references see Ref. [75]).

5.3 Summary

In this chapter, we have introduced our hypothesis of extended quark-lepton complementarity, i.e., all mass ratios and mixings can be expressed in powers of a small expansion parameter $\epsilon \simeq 0.2$ being of the order the Cabibbo angle, where we identify the zeroth order of ϵ with $\pi/4$

5.3 Summary

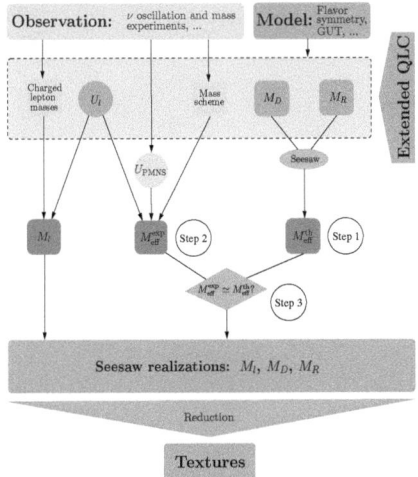

Figure 5.7: Procedure for obtaining seesaw realizations and texture sets in EQLC.

for mixing angles. By using this hypothesis, we have generated all possibilities for the effective neutrino mass matrix $M_{\text{eff}}^{\text{th}}$, which is a product of M_D and M_R as well as all possibilities for $M_{\text{eff}}^{\text{exp}}$, which contains the experimental best-fit values for the PMNS matrix as well as the normal hierarchical neutrino mass spectrum. In the last step, we matched $M_{\text{eff}}^{\text{th}}|_{\epsilon=0.2} \simeq M_{\text{eff}}^{\text{exp}}|_{\epsilon=0.2}$ up to a precision of ϵ^2, which corresponds to the error in oscillation experiments, cf. Fig. 5.7.

Our result was a list of 1981 valid seesaw realizations for which we presented an exemplary list of 72 with corresponding textures [75–77]. We have shown that special cases often considered in literature, such as having a symmetric Dirac mass matrix or small mixing among charged leptons, constitute only a tiny fraction of our possibilities. Our list represents a broad class of *CP* conserving realizations in the sense that large mixings can arise in the charged lepton and the neutrino sector. Moreover, in the neutrino sector, maximal mixing can occur for Dirac and Majorana neutrinos. The mass spectrum of M_D and M_R can be hierarchical or semi-hierarchical which might have direct implications for leptogenesis. All of these mass matrices lead to nearly tribimaximal mixing, *i.e.*, to a very small reactor angle $\theta_{13} \lesssim 1°$ and a deviation from the atmospheric mixing angle testable in future experiments. The seesaw realizations and predictions are very stable under RG effects which can therefore be neglected.

Chapter

6

Lepton Flavor Models

In this chapter, we construct lepton flavor models by direct products of cyclic flavor groups and the Froggatt-Nielsen mechanism as we have done for the $Z_5 \times Z_9$ flavor model of Chap. 4. A machine aided method to predict the seesaw textures of Sec. 5.2.1 is presented. As result, we obtain a list of 22 exemplary flavor models including their seesaw realizations and perform a group space scan [110].

6.1 Flavor Structure

The setup of the models that we want to construct is analogous to the $Z_5 \times Z_9$ flavor model of Chap. 4, i.e., we assume the flavor symmetries to be a direct product of cyclic groups[1]

$$G_F = Z_{n_1} \times Z_{n_2} \times \cdots \times Z_{n_m}, \tag{6.1}$$

where m is the number of Z_n factors and n_k with $k = 1, 2, \ldots, m$ may be different from each other. This is motivated by isomorphisms of groups that can be written as semi-direct products of cyclic groups (see Sec. 3.2), such as $A_4 \sim Z_3 \ltimes (Z_2 \times Z_2)$. The extension to non-Abelian flavor symmetries by semi-direct products will be discussed in Chap. 8. All models are based on the

---

[1]Note, although the flavor symmetries are global, it might be necessary to gauge them in order to survive quantum gravity corrections [111]. The cancellation of anomalies could be achieved by considering extra matter fields.

FN mechanism (Chap. 4), which has a simple scalar sector and the neutrino masses become naturally small due to the type-I seesaw mechanism. Under the flavor symmetry G_F, we assign the following lepton charges for the right-handed charged leptons, the left-handed leptons and the right-handed neutrinos

$$e_i^c \sim (p_1^i, p_2^i, \ldots, p_m^i) = p^i, \quad \ell_i \sim (q_1^i, q_2^i, \ldots, q_m^i) = q^i, \quad \nu_i^c \sim (r_1^i, r_2^i, \ldots, r_m^i) = r^i, \qquad (6.2)$$

where the jth entry in each row vector denotes the Z_{n_j} charge of the particle and i is the generation index. As a convention, we choose the charges for each group Z_{n_k} to be non-negative and lie in the range

$$p_k^i, q_k^i, r_k^i \in \{0, 1, 2, \ldots, n_k - 1\}. \qquad (6.3)$$

For each Z_{n_k}, we assume a flavon that carries charge -1 under Z_{n_k} and is a singlet under all other Z_{n_j} with $j \neq k$ as well as the SM.

6.2 Textures Becoming Models – A Group Space Scan

So far, we have determined the setup of the flavor models, $cf.$, Ref. [110, 112]. However, with an arbitrary choice of flavor charges it would be a bonanza to generate a texture set leading to valid particle masses and mixings. Therefore, we propose a different approach: Since we have constructed in Chap. 5 the largest available set of valid mass matrix textures, we will use this set as our reference set and try to reproduce these textures by flavor symmetries. However, the mass matrices generated in Chap. 5 have no mixings of right-handed charged leptons, $i.e.$, $U_{\ell'}$ is the identity matrix. This was not important at that stage, since $U_{\ell'}$ has no influence on experimental data. However, a flavor symmetry predicts a mass matrix and this usually leads to a non-trivial $U_{\ell'}$. Therefore, we extend our reference set in the following way: we generate the product $M_\ell U_{\ell'}^\dagger$ for each realization of M_ℓ in Chap. 5 by using all possibilities for $U_{\ell'}$ compatible with EQLC, $i.e.$, all mixings are $\pi/4, \epsilon, \epsilon^2$, or 0, and the phases are assumed to be CP conserving. This does not change the phenomenology of the realizations but leads, through introduction of a non-trivial $U_{\ell'}$, to new mass matrices M_ℓ and consequently to a bunch of new realizations. In addition, we allow for M_ℓ a precision of $\mathcal{O}(\epsilon^5)$ in order to account for experimental errors in this sector. This becomes now our new reference set. However, in what follows, we neglect cases of (for this survey) minor interest, $e.g.$, anarchic mass matrices

6.2 Textures Becoming Models – A Group Space Scan

(matrices having only order one entries); see Ref. [110] for more details.

Now, we can systematically scan flavor charges for various flavor symmetries and generate mass matrix textures, which can be compared with the textures in our reference list. If a set is contained in the list, we immediately know that it leads to a viable phenomenology. Contrary, if a matrix set is not contained in our list it does not necessarily mean that the mass matrices are not valid. Only an explicit diagonalization of the mass matrices could check the viability of the model. This shows the advantage of our method being already at the edge of what is nowadays possible with available computer power. As a result, we present in Table 6.1 22 lepton flavor models, *i.e.*, flavor charges with corresponding flavor symmetries and textures. In Table A.2, the corresponding seesaw realizations are shown. These allow a complete reconstruction of the mass matrices including Yukawa couplings.

#	$M_\ell/\langle H \rangle$	$M_D/\langle H \rangle$	M_R/M_{B-L}	p^1, p^2, p^3 q^1, q^2, q^3 r^1, r^2, r^3	G_F
1	$\begin{pmatrix} \epsilon^4 & \epsilon^5 & \epsilon^2 \\ \epsilon^2 & \epsilon^2 & \epsilon^2 \\ \epsilon^2 & \epsilon^4 & 1 \end{pmatrix}$	$\epsilon \begin{pmatrix} \epsilon & \epsilon^2 & \epsilon^2 \\ \epsilon & 1 & \epsilon \\ \epsilon & 1 & \epsilon \end{pmatrix}$	$\epsilon^3 \begin{pmatrix} 1 & \epsilon^2 & 1 \\ \epsilon^2 & 1 & \epsilon^2 \\ 1 & \epsilon^2 & 1 \end{pmatrix}$	$(2,0), (0,0), (2,5)$ $(2,3), (4,1), (3,2)$ $(1,4), (2,6), (0,5)$	$Z_5 \times Z_7$
2	$\epsilon \begin{pmatrix} \epsilon^4 & \epsilon^4 & \epsilon^2 \\ \epsilon^3 & \epsilon^2 & 1 \\ \epsilon^3 & \epsilon^4 & 1 \end{pmatrix}$	$\epsilon \begin{pmatrix} \epsilon & \epsilon^3 & \epsilon \\ \epsilon & 1 & \epsilon^3 \\ \epsilon & \epsilon^2 & \epsilon \end{pmatrix}$	$\epsilon^2 \begin{pmatrix} \epsilon & \epsilon & \epsilon \\ \epsilon & 1 & \epsilon^2 \\ \epsilon & \epsilon^2 & 1 \end{pmatrix}$	$(2,2), (3,2), (2,5)$ $(0,1), (2,2), (4,2)$ $(2,6), (3,4), (1,0)$	$Z_5 \times Z_7$
3	$\epsilon \begin{pmatrix} \epsilon^4 & \epsilon^3 & \epsilon^5 \\ \epsilon^3 & \epsilon^2 & \epsilon^2 \\ \epsilon & \epsilon^2 & 1 \end{pmatrix}$	$\epsilon^2 \begin{pmatrix} \epsilon & \epsilon & \epsilon^3 \\ \epsilon & 1 & 1 \\ \epsilon & 1 & 1 \end{pmatrix}$	$\epsilon \begin{pmatrix} \epsilon & \epsilon & \epsilon^5 \\ \epsilon & 1 & \epsilon^4 \\ \epsilon^5 & \epsilon^4 & 1 \end{pmatrix}$	$(3,7), (3,0), (2,7)$ $(1,5), (3,6), (3,2)$ $(1,4), (2,4), (2,0)$	$Z_5 \times Z_8$

4	$\begin{pmatrix} \epsilon^3 & \epsilon^3 & \epsilon \\ \epsilon & \epsilon^2 & \epsilon^2 & \epsilon^2 \\ \epsilon^2 & \epsilon^4 & 1 \end{pmatrix}$	$\epsilon^3 \begin{pmatrix} \epsilon & \epsilon & \epsilon \\ \epsilon & 1 & 1 \\ \epsilon & 1 & 1 \end{pmatrix}$	$\epsilon \begin{pmatrix} \epsilon & \epsilon^5 & \epsilon \\ \epsilon^5 & 1 & \epsilon^4 \\ \epsilon & \epsilon^4 & 1 \end{pmatrix}$	$(3,0),(0,1),(2,5)$ $(4,2),(3,6),(3,2)$ $(4,0),(3,4),(3,0)$	$Z_5 \times Z_8$
5	$\begin{pmatrix} \epsilon^4 & \epsilon^2 & \epsilon \\ \epsilon & \epsilon^3 & \epsilon^2 & \epsilon^2 \\ \epsilon^5 & \epsilon & 1 \end{pmatrix}$	$\epsilon^2 \begin{pmatrix} \epsilon & \epsilon & \epsilon^3 \\ \epsilon^3 & 1 & \epsilon \\ \epsilon^3 & 1 & \epsilon \end{pmatrix}$	$\epsilon^2 \begin{pmatrix} 1 & \epsilon^3 & 1 \\ \epsilon^3 & 1 & \epsilon^4 \\ 1 & \epsilon^4 & 1 \end{pmatrix}$	$(3,8),(4,3),(0,3)$ $(0,4),(3,7),(4,6)$ $(0,8),(2,4),(1,0)$	$Z_5 \times Z_9$
6	$\begin{pmatrix} \epsilon^4 & \epsilon^2 & \epsilon \\ \epsilon & \epsilon^3 & \epsilon^2 & \epsilon^2 \\ \epsilon^5 & \epsilon & 1 \end{pmatrix}$	$\epsilon^2 \begin{pmatrix} \epsilon^2 & \epsilon^3 & \epsilon \\ \epsilon & 1 & 1 \\ \epsilon & 1 & 1 \end{pmatrix}$	$\epsilon^2 \begin{pmatrix} \epsilon^2 & \epsilon & \epsilon \\ \epsilon & 1 & 1 \\ \epsilon & 1 & 1 \end{pmatrix}$	$(3,6),(2,1),(1,1)$ $(4,6),(1,0),(0,8)$ $(1,8),(0,8),(1,0)$	$Z_5 \times Z_9$
7	$\begin{pmatrix} \epsilon^4 & \epsilon^4 & \epsilon^2 \\ \epsilon & \epsilon^3 & \epsilon^2 & 1 \\ \epsilon^3 & \epsilon^4 & 1 \end{pmatrix}$	$\epsilon^4 \begin{pmatrix} \epsilon^2 & \epsilon & \epsilon^2 \\ \epsilon & \epsilon & \epsilon \\ 1 & \epsilon & 1 \end{pmatrix}$	$\epsilon \begin{pmatrix} 1 & \epsilon^2 & 1 \\ \epsilon^2 & \epsilon & \epsilon^2 \\ 1 & \epsilon^2 & 1 \end{pmatrix}$	$(2,8),(1,8),(1,4)$ $(1,4),(4,4),(3,5)$ $(2,0),(0,1),(2,0)$	$Z_5 \times Z_9$
8	$\begin{pmatrix} \epsilon^3 & \epsilon^4 & \epsilon \\ \epsilon & \epsilon^3 & \epsilon^2 & \epsilon^2 \\ \epsilon^2 & \epsilon^4 & 1 \end{pmatrix}$	$\epsilon^2 \begin{pmatrix} \epsilon^2 & \epsilon & \epsilon^3 \\ \epsilon^2 & 1 & 1 \\ \epsilon^2 & 1 & 1 \end{pmatrix}$	$\epsilon^2 \begin{pmatrix} \epsilon^2 & \epsilon^2 & \epsilon^2 \\ \epsilon^2 & 1 & 1 \\ \epsilon^2 & 1 & 1 \end{pmatrix}$	$(2,1),(1,6),(4,1)$ $(1,6),(0,1),(1,0)$ $(3,6),(0,1),(1,0)$	$Z_5 \times Z_9$
9	$\begin{pmatrix} \epsilon^4 & \epsilon^3 & \epsilon^5 \\ \epsilon^3 & \epsilon^2 & \epsilon^2 \\ \epsilon & \epsilon^4 & 1 \end{pmatrix}$	$\epsilon^3 \begin{pmatrix} \epsilon^2 & \epsilon & \epsilon \\ 1 & 1 & \epsilon^2 \\ 1 & 1 & \epsilon \end{pmatrix}$	$\epsilon^2 \begin{pmatrix} 1 & 1 & \epsilon \\ 1 & 1 & \epsilon \\ \epsilon & \epsilon & 1 \end{pmatrix}$	$(0,2),(2,5),(1,2)$ $(2,3),(4,4),(5,5)$ $(2,0),(3,1),(0,6)$	$Z_6 \times Z_7$

6.2 Textures Becoming Models – A Group Space Scan

10	$\begin{pmatrix} \epsilon^4 & \epsilon^6 & \epsilon^2 \\ \epsilon^2 & \epsilon^2 & \epsilon^2 \\ \epsilon^2 & \epsilon^4 & 1 \end{pmatrix}$	$\epsilon^2 \begin{pmatrix} \epsilon^2 & \epsilon & \epsilon^2 \\ \epsilon^2 & \epsilon & 1 \\ \epsilon^2 & \epsilon & 1 \end{pmatrix}$	$\epsilon \begin{pmatrix} \epsilon^2 & \epsilon^2 & \epsilon^2 \\ \epsilon^2 & \epsilon & \epsilon^4 \\ \epsilon^2 & \epsilon^4 & 1 \end{pmatrix}$	$(2,1), (2,3), (0,1)$ $(1,0), (5,5), (0,6)$ $(4,3), (2,0), (0,3)$	$Z_6 \times Z_7$
11	$\begin{pmatrix} \epsilon^3 & \epsilon^3 & \epsilon^2 \\ \epsilon^2 & \epsilon^2 & \epsilon^2 \\ \epsilon^4 & \epsilon^2 & 1 \end{pmatrix}$	$\epsilon^2 \begin{pmatrix} \epsilon & \epsilon & \epsilon \\ \epsilon & 1 & \epsilon \\ \epsilon & 1 & \epsilon \end{pmatrix}$	$\epsilon \begin{pmatrix} \epsilon & \epsilon & \epsilon^2 \\ \epsilon & 1 & \epsilon^3 \\ \epsilon^2 & \epsilon^3 & 1 \end{pmatrix}$	$(0,0,0), (1,0,2), (1,1,3)$ $(1,1,1), (0,0,2), (1,2,2)$ $(1,0,1), (1,0,2), (0,1,0)$	$Z_2 \times Z_3 \times Z_5$
12	$\begin{pmatrix} \epsilon^4 & \epsilon^5 & \epsilon \\ \epsilon^3 & \epsilon^2 & \epsilon^2 \\ \epsilon^5 & \epsilon^4 & 1 \end{pmatrix}$	$\epsilon^2 \begin{pmatrix} \epsilon & \epsilon & \epsilon^2 \\ \epsilon^2 & 1 & \epsilon^2 \\ \epsilon^2 & 1 & \epsilon \end{pmatrix}$	$\epsilon^2 \begin{pmatrix} 1 & \epsilon & 1 \\ \epsilon & 1 & \epsilon \\ 1 & \epsilon & 1 \end{pmatrix}$	$(1,2,4), (1,1,4), (0,0,2)$ $(0,1,3), (1,0,2), (0,0,3)$ $(0,2,4), (1,0,1), (0,1,0)$	$Z_2 \times Z_4 \times Z_5$
13	$\begin{pmatrix} \epsilon^4 & \epsilon^4 & \epsilon^2 \\ \epsilon^3 & \epsilon^2 & \epsilon^2 \\ \epsilon^5 & \epsilon^4 & 1 \end{pmatrix}$	$\epsilon^2 \begin{pmatrix} \epsilon^3 & \epsilon & \epsilon^2 \\ \epsilon^2 & \epsilon & 1 \\ \epsilon^2 & \epsilon & 1 \end{pmatrix}$	$\begin{pmatrix} \epsilon^2 & \epsilon^3 & \epsilon^2 \\ \epsilon^3 & \epsilon & \epsilon^2 \\ \epsilon^2 & \epsilon^2 & 1 \end{pmatrix}$	$(1,1,1), (1,1,0), (0,3,4)$ $(0,2,2), (0,3,1), (0,1,1)$ $(1,0,1), (0,0,2), (0,0,0)$	$Z_2 \times Z_4 \times Z_5$
14	$\epsilon \begin{pmatrix} \epsilon^4 & \epsilon^5 & \epsilon^2 \\ \epsilon^3 & \epsilon^2 & 1 \\ \epsilon^3 & \epsilon^4 & 1 \end{pmatrix}$	$\epsilon \begin{pmatrix} \epsilon^2 & \epsilon & \epsilon^2 \\ \epsilon^2 & \epsilon & \epsilon^2 \\ 1 & \epsilon^3 & 1 \end{pmatrix}$	$\epsilon^3 \begin{pmatrix} 1 & \epsilon & 1 \\ \epsilon & \epsilon & \epsilon \\ 1 & \epsilon & 1 \end{pmatrix}$	$(0,3), (3,3), (0,6)$ $(3,1), (1,3), (3,3)$ $(1,5), (3,8), (1,5)$	$Z_4 \times Z_9$
15	$\epsilon \begin{pmatrix} \epsilon^4 & \epsilon^4 & \epsilon^2 \\ \epsilon^3 & \epsilon^4 & 1 \\ \epsilon^3 & \epsilon^2 & 1 \end{pmatrix}$	$\epsilon^2 \begin{pmatrix} \epsilon^2 & \epsilon & \epsilon^3 \\ \epsilon^2 & \epsilon & \epsilon^2 \\ \epsilon & \epsilon^2 & 1 \end{pmatrix}$	$\epsilon \begin{pmatrix} \epsilon^2 & \epsilon^2 & \epsilon \\ \epsilon^2 & \epsilon & \epsilon^2 \\ \epsilon & \epsilon^2 & 1 \end{pmatrix}$	$(4,2), (0,2), (4,5)$ $(3,1), (2,2), (0,2)$ $(1,3), (2,3), (0,3)$	$Z_5 \times Z_7$

#	M_1	M_2	M_3	Charges	Group
16	$\begin{pmatrix} \epsilon^4 & \epsilon^3 & \epsilon^2 \\ \epsilon^2 & \epsilon^3 & \epsilon^2 & 1 \\ \epsilon^3 & \epsilon^4 & 1 \end{pmatrix}$	$\epsilon^2 \begin{pmatrix} \epsilon^3 & \epsilon & \epsilon \\ 1 & 1 & \epsilon^3 \\ \epsilon^2 & \epsilon & \epsilon \end{pmatrix}$	$\epsilon^2 \begin{pmatrix} 1 & 1 & \epsilon^3 \\ 1 & 1 & \epsilon^3 \\ \epsilon^3 & \epsilon^3 & 1 \end{pmatrix}$	$(1,6),(0,5),(1,0)$ $(2,7),(0,8),(3,8)$ $(0,8),(4,0),(2,4)$	$Z_5 \times Z_9$
17	$\epsilon \begin{pmatrix} \epsilon^4 & \epsilon^4 & \epsilon^2 \\ \epsilon^3 & \epsilon^4 & 1 \\ \epsilon^3 & \epsilon^2 & 1 \end{pmatrix}$	$\epsilon \begin{pmatrix} \epsilon^2 & \epsilon & \epsilon^3 \\ \epsilon^2 & \epsilon & \epsilon^2 \\ \epsilon^3 & \epsilon^2 & 1 \end{pmatrix}$	$\epsilon^2 \begin{pmatrix} \epsilon^2 & \epsilon^2 & \epsilon^3 \\ \epsilon^2 & \epsilon & \epsilon^2 \\ \epsilon^3 & \epsilon^2 & 1 \end{pmatrix}$	$(4,3),(0,3),(4,0)$ $(3,1),(2,0),(0,0)$ $(2,2),(2,1),(0,1)$	$Z_5 \times Z_7$
18	$\begin{pmatrix} \epsilon^4 & \epsilon^4 & \epsilon^2 \\ \epsilon^3 & \epsilon^2 & \epsilon^2 \\ \epsilon^5 & \epsilon^2 & 1 \end{pmatrix}$	$\epsilon^2 \begin{pmatrix} \epsilon^4 & \epsilon & \epsilon^2 \\ \epsilon^2 & \epsilon & 1 \\ \epsilon^2 & \epsilon & 1 \end{pmatrix}$	$\epsilon \begin{pmatrix} \epsilon^2 & \epsilon^3 & \epsilon^2 \\ \epsilon^3 & \epsilon & \epsilon^5 \\ \epsilon^2 & \epsilon^5 & 1 \end{pmatrix}$	$(4,4),(1,2),(0,1)$ $(1,0),(1,5),(0,6)$ $(2,3),(3,1),(0,3)$	$Z_6 \times Z_7$
19	$\begin{pmatrix} \epsilon^4 & \epsilon^4 & \epsilon^2 \\ \epsilon^2 & \epsilon^2 & \epsilon^2 \\ \epsilon^4 & \epsilon^2 & 1 \end{pmatrix}$	$\epsilon^2 \begin{pmatrix} \epsilon & \epsilon^2 & \epsilon \\ \epsilon & 1 & \epsilon \\ \epsilon & 1 & \epsilon \end{pmatrix}$	$\begin{pmatrix} \epsilon & \epsilon^2 & \epsilon^5 \\ \epsilon^2 & 1 & \epsilon^3 \\ \epsilon^5 & \epsilon^3 & 1 \end{pmatrix}$	$(0,1),(0,3),(4,4)$ $(2,1),(1,4),(1,2)$ $(2,3),(0,3),(0,0)$	$Z_5 \times Z_6$
20	$\begin{pmatrix} \epsilon^4 & \epsilon^5 & \epsilon^2 \\ \epsilon^2 & \epsilon^2 & \epsilon^2 \\ \epsilon^4 & \epsilon^4 & 1 \end{pmatrix}$	$\epsilon \begin{pmatrix} \epsilon^2 & \epsilon & \epsilon^2 \\ \epsilon^2 & \epsilon & 1 \\ \epsilon^2 & \epsilon & 1 \end{pmatrix}$	$\epsilon \begin{pmatrix} \epsilon^2 & \epsilon^2 & \epsilon^2 \\ \epsilon^2 & \epsilon & \epsilon^2 \\ \epsilon^2 & \epsilon^2 & 1 \end{pmatrix}$	$(2,0),(3,5),(1,3)$ $(0,4),(3,2),(4,3)$ $(2,1),(0,4),(2,3)$	$Z_5 \times Z_6$
21	$\begin{pmatrix} \epsilon^4 & \epsilon^5 & \epsilon^2 \\ \epsilon^2 & \epsilon^2 & \epsilon^2 \\ \epsilon^4 & \epsilon^4 & 1 \end{pmatrix}$	$\epsilon \begin{pmatrix} \epsilon & \epsilon^2 & \epsilon \\ \epsilon & 1 & \epsilon \\ \epsilon^3 & 1 & \epsilon \end{pmatrix}$	$\epsilon^2 \begin{pmatrix} \epsilon & \epsilon^2 & \epsilon \\ \epsilon^2 & 1 & \epsilon^3 \\ \epsilon & \epsilon^3 & 1 \end{pmatrix}$	$(3,4),(4,4),(1,2)$ $(3,5),(4,2),(4,4)$ $(2,5),(1,3),(1,0)$	$Z_5 \times Z_6$

6.2 Textures Becoming Models – A Group Space Scan

| 22 | $\begin{pmatrix} \epsilon^4 & \epsilon^3 & \epsilon^2 \\ \epsilon^2 & \epsilon^2 & \epsilon^3 \\ \epsilon^5 & \epsilon & 1 \end{pmatrix}$ | $\begin{pmatrix} \epsilon^2 & \epsilon & \epsilon^2 \\ 1 & \epsilon & 1 \\ 1 & \epsilon^3 & 1 \end{pmatrix}$ | $\begin{pmatrix} 1 & \epsilon^3 & 1 \\ \epsilon^3 & \epsilon & \epsilon^3 \\ 1 & \epsilon^3 & 1 \end{pmatrix}$ | $(2,6),(0,0),(0,1)$ $(0,6),(1,1),(0,8)$ $(1,0),(2,5),(1,0)$ | $Z_3 \times Z_9$ |

Table 6.1: List of 22 valid lepton flavor models for nearly tribimaximal mixing with their explicit flavor charges p^i, q^i, and r^i, under the flavor symmetry G_f and the resulting textures.

All 22 models of Tables 6.1 and A.2 lead to the following PMNS mixing angles

$$34° \lesssim \theta_{12} \lesssim 39°, \quad \theta_{13} \lesssim 1°, \quad \theta_{23} \approx 52°, \tag{6.4}$$

which are in agreement with neutrino oscillation data (at 3σ CL). Moreover, the significant deviation of the atmospheric mixing angle from maximal makes the models testable at future experiments. Note, one seesaw realization can, in principle, be generated by more than one flavor symmetry. In Fig. 6.1, we show the full extent of our group space scan, namely 6021 lepton flavor models leading to 2093 distinct texture sets.[2]

In Fig. 6.1, we can see the trend, up to periodic fluctuations, that an increasing group order leads to more lepton flavor models, as expected. Therefore, we give a rough estimate of the group order which might be needed in order to predict any given texture set. Our CP conserving mass matrices have entries up to ϵ^2 for M_D and M_R, and ϵ^4 for M_ℓ (higher orders are approximated by 0). This leads to $4^{6+9} \cdot 6^9$ possible texture sets including phases 0 and π. These should be producible by products of cyclic groups with 9 different charges per factor, i.e., we have $|G_f|^9$ possible charge assignments. In order to possibly produce some random texture set, the number of charge assignments should exceed the number possible texture sets, i.e., $|G_f| \gtrsim 4^{\frac{15}{9}} \cdot 6 \simeq 60$. This means that, for example, the group $Z_2 \times Z_3 \times Z_4 \times Z_5$ should be able to produce any of our texture sets.

---

[2]We restrict ourselves for simplicity to groups up to order 40 consisting of one Z_n, 45 for a direct product of two Z_n, 30 for three factors, and 24 for four factors, with $n \leq 9$ for more than one factor.

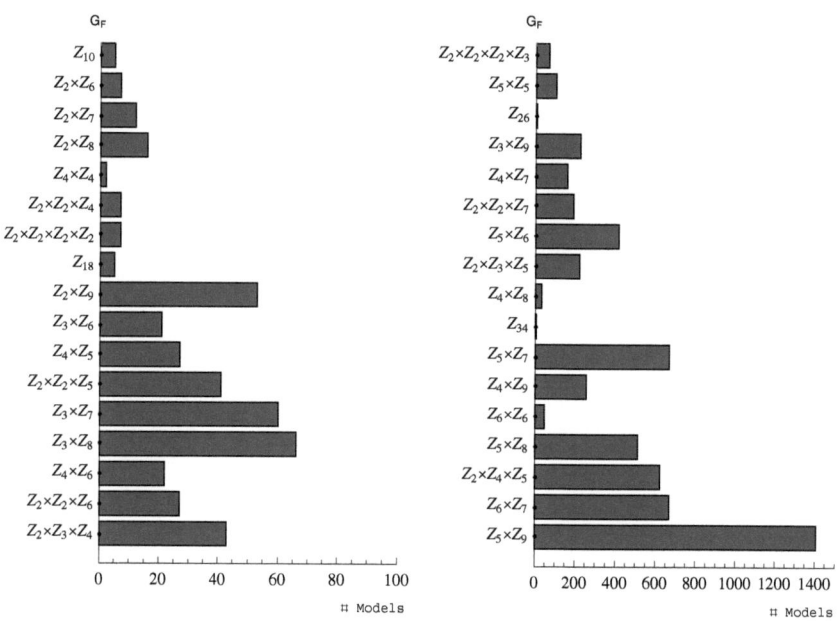

Figure 6.1: Number of flavor models leading to nearly tribimaximal lepton mixing as a function of the flavor group G_F for increasing group order. In the left (right) panel we have $10 \leq |G_F| \leq 24$ ($24 \leq |G_F| \leq 45$).

6.3 Summary

In this chapter, we have presented a model setup based on flavor symmetries of direct products of cyclic groups Z_n and the Froggatt-Nielsen mechanism. A method how the seesaw textures obtained in Sec. 5.2.1 can be generated by these flavor groups was introduced. We have systematically scanned all possible flavor charge assignments for leptons in order to compare the resulting mass matrices with our textures. If they match, we immediately know, without the need for diagonalization, that the model leads to viable masses and mixings. As a result, we have found a list of 22 example models including their seesaw realizations, as well as an overview of 6 021 lepton flavor models for varying flavor groups [110].

Chapter

7

Lepton Flavor Violation

In this chapter, we survey the lepton flavor violation (LFV) decay rates $\text{Br}(\mu \to e\gamma)$, $\text{Br}(\tau \to \mu\gamma)$, and $\text{Br}(\tau \to e\gamma)$ for the LHC relevant scenario SPS1a' in minimal supergravity (mSUGRA) for a broad class of non-trivial lepton mass matrix textures shown in Tables 5.1 and A.1 (see Ref. [113]). This is done for the most general CP violating forms of the textures. Moreover, the correlation between LFV rates and low energy lepton mixing parameters is investigated.

7.1 Charged LFV in SUSY

In the SM, no LFV is present. Even if the SM is amended by heavy right-handed neutrinos ν^c the LFV rates are tiny due to the suppression of their mass scale. This may change completely in SUSY seesaw models considered here. The reason is that the ν^c influence the mixing of the sleptons and the trilinear coupling in the Minimal Supersymmetric Standard Model (MSSM) via radiative corrections. This occurs mainly via intermediate left-handed slepton flavor transitions. The most important low energy processes are the rare decays $l_i \to l_j \gamma$, where $i \neq j \in e, \mu, \tau$. The current bounds on these processes, which provide by now the most stringent bounds on LFV in the SUSY seesaw models, as well as the expected future sensitivities, are shown in Table 7.1.

	Br($\mu \to e\gamma$)	Br($\tau \to \mu\gamma$)	Br($\tau \to e\gamma$)
Present	1.2×10^{-11} [114]	6.8×10^{-8} [115]	1.1×10^{-7}
Expected	10^{-13}	$\approx 10^{-8}$	$\approx 10^{-8}$

Table 7.1: Current bounds and expected future sensitivities of direct experimental LFV searches.

We denote the superpotential as

$$W = -(Y_\ell)_{ij} e_i^c \ell_j H_1 - (Y_D)_{ij} \nu_i^c \ell_j H_2 + \frac{1}{2}(M_R)_{ij} \nu_i^c \nu_j^c, \qquad (7.1)$$

where ℓ_i, e_i^c, and ν_i^c, are the matter superfields, and H_1 and H_2 the usual Higgs superfield doublets generating the down- (H_1) and up-type masses (H_2). Y_ℓ, Y_D, and M_R are the charged lepton and Dirac Yukawa coupling matrix and the Majorana mass matrix, respectively. After electroweak symmetry breaking, the Higgs doublets develop the VEVs $\langle H_i^0 \rangle$, where $\langle H_2^0 \rangle = v \sin\beta$, with $v = 174\,\text{GeV}$ and $\tan\beta = \langle H_2^0 \rangle / \langle H_1^0 \rangle$.

In the SUSY-breaking scenario mSUGRA, the 6×6 slepton mass matrix does not contain flavor mixing terms and may be written as

$$m_{\tilde{\ell}}^2 = \begin{pmatrix} m_L^2 & m_{LR}^{2\dagger} \\ m_{LR}^2 & m_R^2 \end{pmatrix}_{\text{MSSM}} + \begin{pmatrix} \delta m_L^2 & \delta m_{LR}^{2\dagger} \\ \delta m_{LR}^2 & \delta m_R^2 \end{pmatrix}_{\nu^c}, \qquad (7.2)$$

where the first summand is the usual MSSM mass matrix without right-handed neutrinos and the second one parameterizes the corrections by ν^c and can, to leading logarithmic approximation, be written as [116],

$$\begin{aligned} \delta m_L^2 &= -\tfrac{1}{8\pi^2}(3m_0^2 + A_0^2) Y_D^\dagger L Y_D, \\ \delta m_R^2 &= 0, \\ \delta m_{LR}^2 &= -\tfrac{3}{16\pi^2} A_0 v \cos\beta Y_\ell Y_D^\dagger L Y_D, \end{aligned} \qquad (7.3)$$

where $L_{ij} = \ln(M_X/m_i^R)\delta_{ij}$, Y_ℓ and Y_D are the Yukawa coupling matrices of Eq. (7.1), m_i^R are the heavy neutrino masses, and m_0 and A_0 are the universal scalar mass and trilinear coupling, respectively, all at the GUT scale M_X. These flavor off-diagonal virtual effects lead to charged LFV (see Ref. [117] and references therein for more details) and suppress a given process relative to the flavor conserving one by a small factor $|(\delta m_L^2)_{ij}/\widetilde{m}^2|^2$ ($i \neq j$), where $(\delta m_L^2)_{ij}$ are the off-diagonal elements of Eq. (7.3) and $\widetilde{m}^2$ is of the order the relevant sparticle

7.2 LFV for CP Conserving and CP Violating Textures

masses in the loops involved in the process. To leading order in the LFV couplings [116], the branching ratios (BR) for the rare decays are

$$\text{Br}(l_i \to l_j \gamma) \propto \alpha^3 m_{l_i}^5 \frac{|(\delta m_L^2)_{ij}|^2}{\widetilde{m}^8} \tan^2 \beta. \tag{7.4}$$

Note that this expression serves only as an illustration. For the numerical calculations, we use the full one loop result for $\text{Br}(l_i \to l_j\gamma)$, as outlined in [117]. As a reference point, we take the LHC relevant mSUGRA benchmark scenario SPS1a', i.e., the parameters used to calculate the LFV rates are $M_X = 2.5 \times 10^{16}$ GeV for the GUT scale, $m_\nu = 5 \times 10^{-2}$ GeV for the effective neutrino mass scale with normal ordering, a universal gaugino mass $m_{1/2} = 250$ GeV and a universal scalar mass $m_0 = 70$ GeV at the GUT scale, $\tan\beta = 10$, a positive sign of the Higgs mixing parameter μ, and a universal trilinear coupling parameter $A_0 = -300$.

7.2 LFV for CP Conserving and CP Violating Textures

In this section, we investigate the LFV rates for the seesaw realizations[1] in Tables 5.1 and A.1 for the mSUGRA benchmark scenario SPS1a' given in Sec. 7.1. For a unified picture, we will go to the basis where the charged leptons and heavy right-handed neutrinos are diagonal, i.e., the PMNS matrix now diagonalizes the effective neutrino mass matrix. This amounts just to a rotation of the flavor basis and has therefore no influence on observables. By using the notation of Sec. 4.2.1, M_D reads

$$M_D = K_R^* \widehat{U}_R^\dagger \widetilde{D} \widehat{U}_{D'}^* M_D^{\text{diag}} \widetilde{K} \widehat{U}_D^T D_D U_\ell^*. \tag{7.5}$$

A normalization of the heaviest eigenvalue of M_{eff} and M_R to one, i.e., m_3 and m_3^R are factored out of the mass matrices, also changes $M_D \to \sqrt{\frac{m_\nu}{m_3} \frac{M_{B-L}}{m_3^R}} M_D = \langle H_2^0 \rangle Y_D$ in order to keep the seesaw formula invariant. This leads to

$$Y_D = \frac{1}{\langle H_2^0 \rangle} \sqrt{\frac{m_\nu}{m_3} \frac{M_{B-L}}{m_3^R}} K_R^* \widehat{U}_R^\dagger \widetilde{D} \widehat{U}_{D'}^* M_D^{\text{diag}} \widetilde{K} \widehat{U}_D^T D_D U_\ell^* . \tag{7.6}$$

---

[1]Note, in Chap. 6 we have demonstrated that each texture set/realization can, in principal, be obtained from a flavor symmetry.

Usually one parameterizes Y_D [118] as[2]

$$Y_D = \frac{1}{v \sin \beta} \sqrt{M_R^{\text{diag}}} \cdot R \cdot \sqrt{M_{\text{eff}}^{\text{diag}}} \cdot U_{\text{PMNS}}^T , \qquad (7.7)$$

where, R is a complex orthogonal matrix, which can be written in terms of 3 complex angles $\theta_i = x_i + i y_i$ as

$$R = \begin{pmatrix} c_2 c_3 & -c_1 s_3 - s_1 s_2 c_3 & s_1 s_3 - c_1 s_2 c_3 \\ c_2 s_3 & c_1 c_3 - s_1 s_2 s_3 & -s_1 c_3 - c_1 s_2 s_3 \\ s_2 & s_1 c_2 & c_1 c_2 \end{pmatrix}, \qquad (7.8)$$

with $(c_i, s_i) = (\cos \theta_i, \sin \theta_i) = (\cos x_i \cosh y_i - i \sin x_i \sinh y_i, \sin x_i \cosh y_i + i \cos x_i \sinh y_i)$. The parameters can take the values $x_i \in [0, 2\pi[$ and $y_i \in]-\infty, \infty[$. However, in a perturbative theory, the y_i are constrained to values $|y_i| \lesssim \mathcal{O}(1)$. The parameterization of Eq. (7.7) has the advantage that by inserting the experimental values for neutrino masses and mixings, a scan of Y_D yields always a valid low energy phenomenology. However, the exact connection to models, e.g., flavor models, has been lost after rotating to the special basis where M_ℓ and M_R are diagonal.

Therefore, we use, in what follows, the parameterization of Eq. (7.5) to calculate the LFV rates of Eq. (7.4) for the seesaw realizations in Tables 5.1 and A.1. Albeit these realizations exhibit a perfect fit to experimental data, they are CP conserving by construction. Therefore, we generalize our realizations to cases with CP violation. For this we provide each Yukawa coupling of M_ℓ, M_D, and M_R with an a priori unconstrained phase factor. This leads to a complexification of the seesaw realizations with free phases. These phases certainly have influence on the mass and mixing parameters of the mass matrices. Therefore, we demand the PMNS mixing angles θ_{12} and θ_{23} to be within their current 1σ bound (see Sec. 5.2.2), $\theta_{13} < 5°$, and the mass eigenvalues of each realization should not vary by a factor larger than 1.5. In addition, to ensure the perturbativity of the Higgs sector, we require $|Y_{D3}|^2/(4\pi) < 0.3$, where Y_{D3} is the largest eigenvalue of $Y_D^\dagger Y_D$. In this way, we have assured the phenomenological viability of each complexified texture set. The results of our analysis will be presented in the next section.

---

[2]The definition of U_{PMNS} in [118] differs from our definition by a complex conjugation.

7.3 Results for LFV-Rates

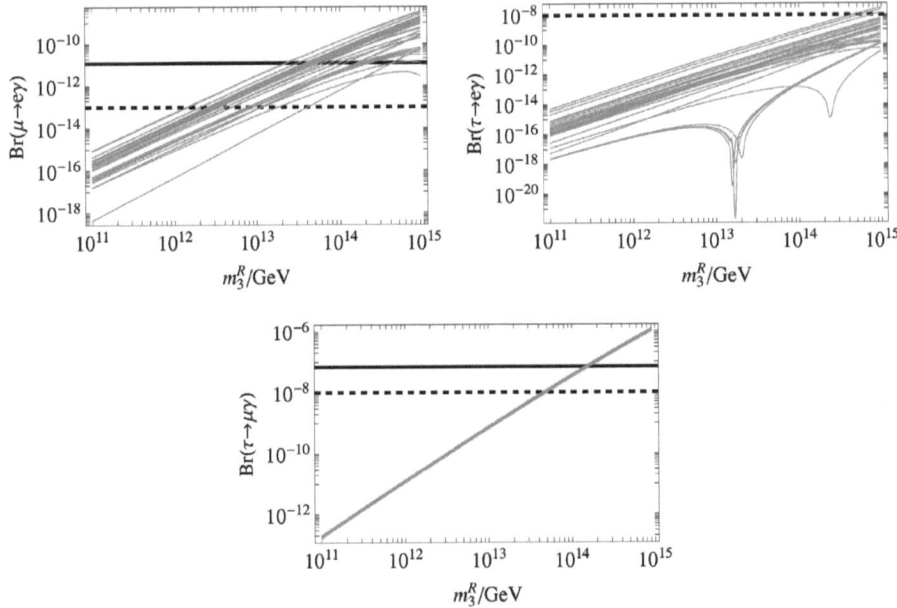

Figure 7.1: $\text{Br}(\mu \to e\gamma)$ (top left), $\text{Br}(\tau \to e\gamma)$ (top right), and $\text{Br}(\tau \to \mu\gamma)$ (bottom), for the reference list of 72 CP conserving seesaw realizations for the mSUGRA scenario SPS1a'. The solid (dashed) line represents the current (future) experimental bound.

7.3 Results for LFV-Rates

In this section, we present the actual results of our LFV analysis of the 72 seesaw realizations in Tables 5.1 and A.1. For the CP conserving case, the rare decay rates $\text{Br}(\mu \to e\gamma)$, $\text{Br}(\tau \to \mu\gamma)$, and $\text{Br}(\tau \to e\gamma)$, as functions of the heaviest right-handed Majorana neutrino mass m_3^R are plotted in Fig. 7.1. The poles in the graphs for $\text{Br}(\tau \to e\gamma)$ are due to an accidental cancellation of the chargino and neutralino amplitude. For the 72 real textures, $\text{Br}(\tau \to \mu\gamma)$ varies hardly for a fixed value of m_3^R compared to $\text{Br}(\mu \to e\gamma)$, which shows a variation of at least two orders of magnitude and $\text{Br}(\tau \to e\gamma)$, with at least three orders of magnitude. This would allow a differentiation between the realizations and m_3^R, respectively. For $\tau \to \mu\gamma$, a future non-observation would imply m_3^R to be less than about 4×10^{13} GeV. The rates for $\tau \to e\gamma$ are practically not touched even by the upcoming PSI experiment. However, the most restrictive bound is given by $\text{Br}(\mu \to e\gamma)$, i.e., for all realizations to be valid, m_3^R has to be smaller than

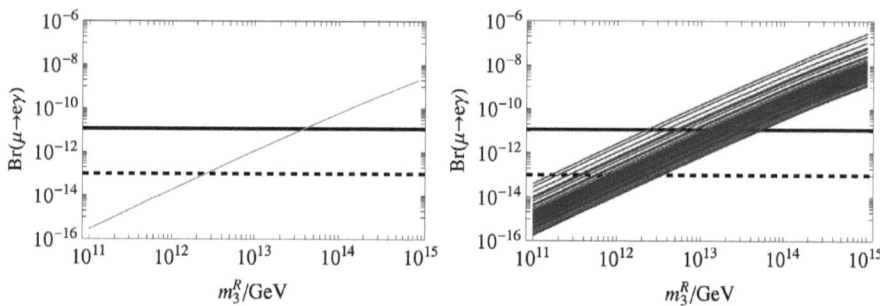

Figure 7.2: Br($\mu \to e\gamma$) as a function of m_3^R for the mSUGRA scenario SPS1a' and the CP conserving seesaw realization #1 (left) of Table 5.1 and 60 random complexifications (right) of the same texture set. The solid (dashed) line represents the current (future) experimental bound.

about 10^{13} GeV, implying the current experimental bound of 10^{-11}. For the expected future sensitivity of 10^{-13}, the upper limit of m_3^R, valid for all realizations, would be 10^{12} GeV which would let the seesaw mechanism be disputable in this scenario.

For the complexified texture sets, the situation changes even though they are generalizations of the CP conserving textures we analyzed before. As an example, we show in Fig. 7.2 Br($\mu \to e\gamma$) as a function of m_3^R for the first seesaw realization of Table 5.1.[3] The left plot refers to the CP conserving mass matrix set while the right plot illustrates the repercussion of 60 complexifications of the same texture set #1. This means, the complexification erodes the significance of the branching ratio, compared to the CP conserving case, since, for a fixed m_3^R, it is spread by a factor of typically about 10. This (mainly) increasing of Br($\mu \to e\gamma$) by the presence of phases is a general feature for all 72 considered realizations, which means that we cannot differentiate anymore between the underlying texture sets while uncontrolled phases are present. For CP conserving seesaw models it is known that a huge dependence of Br($\mu \to e\gamma$) on the PMNS mixing angle θ_{13} exists. This is due to the accidental cancellation of the chargino and neutralino amplitude. Therefore, in Fig. 7.3 we show for 500 complexifications of the texture set #1 of Table 5.1 the correlation of Br($\mu \to e\gamma$) with the PMNS mixing angles θ_{12}, θ_{13}, and θ_{23} and the Dirac CP phase.[4] However, Fig. 7.3 do not show any dependence of Br($\mu \to e\gamma$) from low energy observables at the presence of CP violation. This means, as long

---

[3] The electric dipole moments for the complex texture sets are several orders of magnitude below the current experimental bound as general expected in seesaw models.

[4] For Majorana phases no dependencies are expected.

7.4 Summary

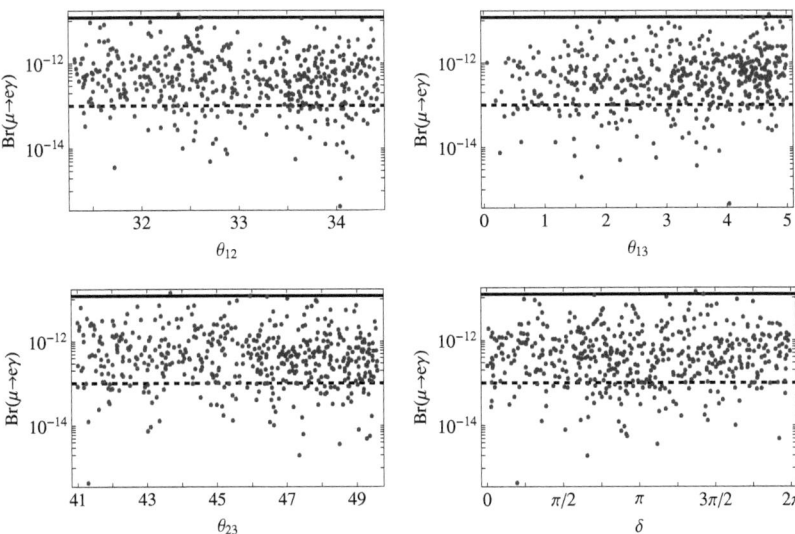

Figure 7.3: Br($\mu \to e\gamma$) as function of the PMNS mixing angles θ_{12}, θ_{13}, and θ_{23} and the Dirac CP phase for 500 complexifications of the seesaw realization #1 in the mSUGRA scenario SPS1a'. The solid (dashed) lines represent the current (future) experimental bounds at $M_R = 2.5 \times 10^{12}$ GeV.

as a model does not control the phases, a differentiation by low energy observables seems to be impossible.

7.4 Summary

In this chapter, we have studied the branching ratios Br($\mu \to e\gamma$), Br($\tau \to \mu\gamma$), and Br($\tau \to e\gamma$), in mSUGRA at the Lagrangian level for the broad class of non-trivial lepton mass matrix textures in Tables 5.1 and A.1 (see Ref. [113]). This has been done for the most general CP violating forms of the textures. For these, we have demonstrated that the branching ratios are enhanced by typically an order of magnitude as compared to the CP conserving case. Moreover, the branching ratios exhibit a strong dependence on the choice of the phases in the Lagrangian, *i.e.*, the LFV rates appear to be essentially uncorrelated with the possible high- and low energy lepton mixing parameters as long as the model does not control the phases.

Chapter 8

5D SUSY $SU(5)$ GUTS with Non-Abelian Flavor Symmetries

In this chapter, we extend the procedure of Chap. 6 in various ways to obtain 5D SUSY $SU(5)$ GUTS with non-Abelian flavor symmetries [112]. For this, we complete each seesaw representation with the corresponding $SU(5)$ compatible quark mass matrices. As a result, we obtain 437 $SU(5)$ GUTs with Abelian flavor symmetries. Each yields an excellent fit to all fermion masses, the CKM matrix and nearly tribimaximal lepton mixing. We give an explicit model example how the Abelian flavor symmetries can be extended to non-Abelian groups, demonstrate the impact of the non-Abelian nature of the flavor symmetry on the atmospheric mixing angle, and present an embedding into a 5D scenario.

8.1 4D SUSY $SU(5)$ GUTs with Abelian Flavor Symmetries

The approach to construct 4D SUSY $SU(5)$ GUTs with Abelian flavor symmetries is analogous to Chap. 6. The models are defined by flavor symmetries which are direct products of cyclic groups Z_n. However, this time the fields are supposed to be supermultiplets of a $SU(5)$ GUT, *i.e.*, the particles will be embedded in the following $SU(5)$ representations

$$\begin{aligned} \mathbf{10} &= (\overline{\mathbf{3}},\mathbf{1})_{-\frac{2}{3}} + (\mathbf{3},\mathbf{2})_{\frac{1}{6}} + (\mathbf{1},\mathbf{1})_1 & (u^c, q, e^c), \\ \overline{\mathbf{5}} &= (\overline{\mathbf{3}},\mathbf{1})_{\frac{1}{3}} + (\mathbf{1},\overline{\mathbf{2}})_{-\frac{1}{2}} & (d^c, \ell), \\ \mathbf{1} &= (\overline{\mathbf{1}},\mathbf{1})_0 & (\nu^c). \end{aligned} \qquad (8.1)$$

The Yukawa coupling Langrangian takes then the form

$$\mathcal{L}_Y^{SU(5)} = \int d^2\theta \left[Y_{ij}^u \mathbf{10}_i \mathbf{10}_j \mathbf{5}^H + Y_{ij}^d \mathbf{10}_i \overline{\mathbf{5}}_j \overline{\mathbf{5}}^H + Y_{ij}^D \overline{\mathbf{5}}_i \mathbf{1}_j \mathbf{5}^H + M_R Y_{ij}^R \mathbf{1}_i \mathbf{1}_j + \text{H.c.} \right], \qquad (8.2)$$

where $\mathbf{5}^H$ and $\overline{\mathbf{5}}^H$ are the Higgs multiplets. This means, due to unification, $M_d = M_\ell^T$. Thus, not only the mass ratios are equal but also the absolute mass scales, *i.e.*,

$$m_b = m_\tau, \qquad m_s = m_\mu, \qquad m_d = m_e. \qquad (8.3)$$

Moreover, the mass matrices in the general parameterization

$$\begin{aligned} M_u &= V_u M_u^{\text{diag}} V_{u'}^\dagger, & M_d &= V_d M_d^{\text{diag}} V_{d'}^\dagger, \\ M_D &= U_D M_D^{\text{diag}} U_{D'}^\dagger, & M_R &= U_R M_R^{\text{diag}} U_R^T, \end{aligned} \qquad (8.4)$$

can in $SU(5)$ GUTs, by using our notation of Sec. 4.2.1 as well as $V_d = U_{\ell'}^*$, $V_{d'} = U_\ell^*$, and $V_{\text{CKM}} = V_d^\dagger V_u$, be rewritten as

$$\begin{aligned} M_u &= U_{\ell'}^* V_{\text{CKM}} M_u^{\text{diag}} V_{\text{CKM}}^\dagger U_{\ell'}^T, & M_d &= M_\ell^T, \\ M_D &= U_D M_D^{\text{diag}} U_{D'}^\dagger, & M_R &= U_R M_R^{\text{diag}} U_R^T. \end{aligned} \qquad (8.5)$$

Using this setup we can now follow the guideline of Chap. 6. We assign under a flavor symmetry G_A, which is a direct product of cyclic groups, the following charges:

$$\begin{aligned} \mathbf{10}_i &\sim (p_1^i, p_2^i, \ldots, p_m^i) = p^i, \\ \overline{\mathbf{5}}_i &\sim (q_1^i, q_2^i, \ldots, q_m^i) = q^i, \\ \mathbf{1}_i &\sim (r_1^i, r_2^i, \ldots, r_m^i) = r^i, \end{aligned} \qquad (8.6)$$

where i is the generation index and m denotes the number of Z_n factors of G_A.

8.1 4D SUSY $SU(5)$ GUTs with Abelian Flavor Symmetries

As an explicit example, we consider the group $G_A = Z_3 \times Z_8 \times Z_9$ with charge assignments

$$
\begin{aligned}
&10_1 \sim (1,1,6), \quad 10_2 \sim (0,3,1), \quad 10_3 \sim (0,0,0)\,, \\
&\bar{5}_1 \sim (1,4,2), \quad \bar{5}_2 \sim (0,7,0), \quad \bar{5}_3 \sim (0,0,1)\,, \\
&1_1 \sim (2,0,6), \quad 1_2 \sim (2,6,0), \quad 1_3 \sim (2,0,6)\,.
\end{aligned}
\tag{8.7}
$$

These flavor charges generate via the FN mechanism, cf. Chap. 6, the mass matrix textures

$$
Y_{ij}^u \sim \begin{pmatrix} \epsilon^6 & \epsilon^7 & \epsilon^5 \\ \epsilon^7 & \epsilon^4 & \epsilon^4 \\ \epsilon^5 & \epsilon^4 & 1 \end{pmatrix}, \quad Y_{ij}^d \sim \epsilon \begin{pmatrix} \epsilon^4 & \epsilon^3 & \epsilon^3 \\ \epsilon^4 & \epsilon^2 & \epsilon^4 \\ \epsilon^6 & 1 & 1 \end{pmatrix},
\tag{8.8a}
$$

$$
Y_{ij}^D \sim \epsilon^3 \begin{pmatrix} \epsilon^2 & \epsilon & \epsilon^2 \\ \epsilon^2 & \epsilon & \epsilon^2 \\ 1 & \epsilon & 1 \end{pmatrix}, \quad Y_{ij}^R \sim \epsilon^4 \begin{pmatrix} 1 & \epsilon^2 & 1 \\ \epsilon^2 & \epsilon & \epsilon^2 \\ 1 & \epsilon^2 & 1 \end{pmatrix}.
\tag{8.8b}
$$

Since this set of lepton textures is contained in the seesaw realizations, constructed in Sec. 6.2 and because the quark matrices obey Eq. (8.5), we do not only know that these textures lead to viable lepton masses and mixings but we can also reconstruct the corresponding Yukawa couplings. This is also true for the quarks by using the mass ratios of Eq. (4.1) and the CKM mixings of Eq. (4.4) for Eq. (8.5). The reconstructed mass matrices including their Yukawa couplings are

$$
Y^u \sim \begin{pmatrix} \epsilon^6 & 0 & -\frac{\epsilon^5}{3} \\ 0 & \epsilon^4 & -\frac{\epsilon^4}{2}+\frac{\epsilon^6}{8} \\ -\frac{\epsilon^5}{3} & -\frac{\epsilon^4}{2}+\frac{\epsilon^6}{8} & 1 \end{pmatrix}, \quad Y^d \sim \begin{pmatrix} -\epsilon^4 & \frac{\epsilon^3}{\sqrt{2}} & -\frac{\epsilon^3}{\sqrt{2}} \\ \epsilon^4 & \sqrt{2}\epsilon^2 - \frac{\epsilon^4}{2\sqrt{2}} & \frac{\epsilon^4}{2\sqrt{2}} \\ 0 & \frac{1}{\sqrt{2}}-\frac{3\epsilon^4}{2\sqrt{2}} & \frac{1}{\sqrt{2}}+\frac{\epsilon^4}{2\sqrt{2}} \end{pmatrix},
\tag{8.9a}
$$

$$
Y^D \sim \begin{pmatrix} (-\frac{1}{2}-\frac{1}{\sqrt{2}})\epsilon^2 & \frac{\epsilon}{\sqrt{2}} & (\frac{1}{2}-\sqrt{2})\epsilon^2 \\ (\frac{1}{2}+\frac{1}{\sqrt{2}})\epsilon^2 & \frac{\epsilon}{\sqrt{2}} & -\frac{\epsilon^2}{2} \\ -\frac{1}{\sqrt{2}} & -\frac{\epsilon}{\sqrt{2}} & -\frac{1}{\sqrt{2}}+\frac{\epsilon^2}{2\sqrt{2}} \end{pmatrix}, \quad Y^R \sim \begin{pmatrix} \frac{1}{2}+\frac{\epsilon^2}{2} & \frac{\epsilon^2}{2} & \frac{1}{2}-\frac{\epsilon^2}{2} \\ \frac{\epsilon^2}{2} & \epsilon & \frac{\epsilon^2}{2} \\ \frac{1}{2}-\frac{\epsilon^2}{2} & \frac{\epsilon^2}{2} & \frac{1}{2}+\frac{\epsilon^2}{2} \end{pmatrix}.
\tag{8.9b}
$$

This leads to a hierarchical mass spectrum for $M_D^{\text{diag}} \sim \text{diag}(\epsilon^2, \epsilon, 1)$, $M_R^{\text{diag}} \sim \text{diag}(\epsilon^2, \epsilon, 1)$, and for the effective neutrino masses $M_\nu^{\text{diag}} \sim \text{diag}(\epsilon^2, \epsilon, 1)$, as well as to the mixings and phases

$$\begin{aligned}
(\theta_{12}^\ell, \theta_{13}^\ell, \theta_{23}^\ell, \delta^\ell, \alpha_1^\ell, \alpha_2^\ell) &= (\epsilon^2, 0, \tfrac{\pi}{4}, 0, \pi, \pi) \,, \\
(\theta_{12}^{\ell'}, \theta_{13}^{\ell'}, \theta_{23}^{\ell'}, \delta^{\ell'}, \alpha_1^{\ell'}, \alpha_2^{\ell'}) &= (\epsilon, 0, \epsilon^2, 0, 0, \pi) \,, \\
(\theta_{12}^D, \theta_{13}^D, \theta_{23}^D, \delta^D, \varphi_1^D, \varphi_2^D, \varphi_3^D) &= (\tfrac{\pi}{4}, \epsilon^2, \epsilon^2, \pi, 0, 0, \pi) \,, \\
(\theta_{12}^{D'}, \theta_{13}^{D'}, \theta_{23}^{D'}, \delta^{D'}, \alpha_1^{D'}, \alpha_2^{D'}) &= (\epsilon^2, \tfrac{\pi}{4}, \epsilon, 0, \pi, 0) \,, \\
(\theta_{12}^R, \theta_{13}^R, \theta_{23}^R, \delta^R, \varphi_1^R, \varphi_2^R, \varphi_3^R) &= (\epsilon^2, \tfrac{\pi}{4}, \epsilon^2, \pi, 0, \pi, \pi) \,,
\end{aligned} \quad (8.10)$$

and we choose $\varphi_i^\ell = \varphi_i^{\ell'} = \varphi_i^{D'} = \alpha_j^D = 0$ for $i = 1, 2, 3$ and $j = 1, 2$ since these phases appear only in combination with other phases (*cf.*, Sec. 4.2.1). Furthermore, our model has $\tan \beta \sim 10$ and a realistic mass ratio between the first two charged fermion generations may be achieved by the Georgi-Jarlskog mechanism [60]. The PMNS mixings $\theta_{12} \approx 34°$, $\theta_{13} \approx 0.2°$, and $\theta_{23} \approx 52°$, exhibit nearly tribimaximal lepton mixing. Note that the quark mixings are a priori in perfect agreement with experiment since we used them in Eq. (8.5) for the mass matrix construction. Although the flavor symmetry is supposed to be broken at some high scale, we can neglect modifications by RG effects within our precision (see Sec. 5.2.2).

8.2 Scanning SUSY $SU(5)$ GUTs with Abelian Flavor Symmetries

The $SU(5)$ model presented in Sec. 8.1 illustrates the method as well as the goal that we want to reach in this section: a systematic scan of $SU(5)$ models with Abelian flavor symmetries, analogous to the lepton flavor model scan of Sec. 6.2. As a starting point, we take again advantage of the reference list (of seesaw realizations) described in Sec. 6.2. However, we amend each realization by its $SU(5)$ compatible up- and down-quark mass matrix according to Eq. (8.5). In this way, we obtain a $SU(5)$ compatible set of textures M_u, M_d, M_ℓ, M_D, and M_R, which we will take as our new reference list modulo trivial realizations containing, *e.g.*, anarchic textures. Again, we generate all possible flavor charges for each considered flavor symmetry G_A being a direct-product of cyclic groups. The resulting scan with 437 viable SUSY $SU(5)$ GUTs is shown in Fig. 8.1. Each model yields an excellent fit to quark and lepton masses, V_{CKM}, and nearly tribimaximal lepton mixing, *i.e.*, $\theta_{13} \ll 1°$ and a normal neutrino mass spectrum. Note that the number of models for a certain flavor symmetry increases, up to some modulation,

8.3 5D SUSY $SU(5)$ GUTs with Non-Abelian Flavor Symmetries

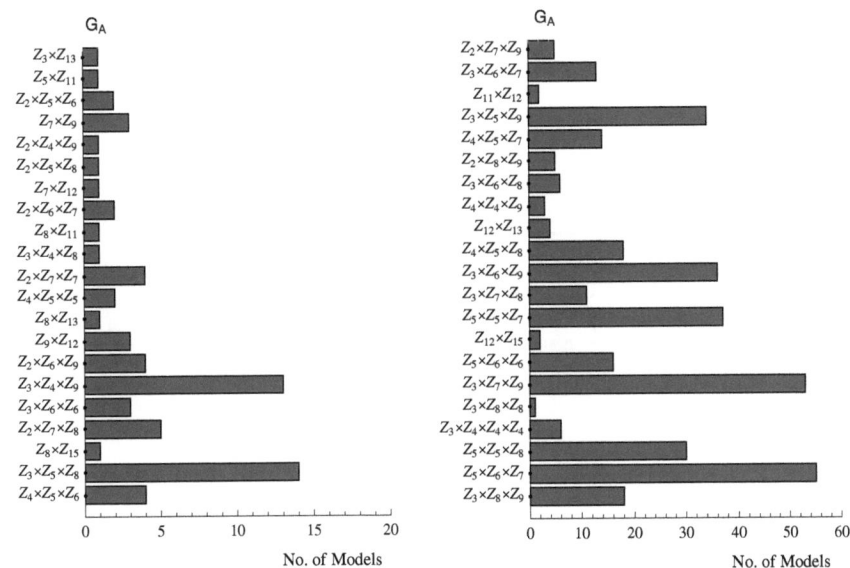

Figure 8.1: Overview of $SU(5) \times G_A$ models for varying flavor group G_A. The graph summarizes 437 realistic GUTs.

with its group order. The simplest flavor group in our scan, *i.e.*, $Z_3 \times Z_{13}$, generates two $SU(5)$ GUTs and has group order 39. The most "efficient" group is the second largest one, *i.e.*, $Z_5 \times Z_6 \times Z_7$. It provides 55 models and has group order 210.

8.3 5D SUSY $SU(5)$ GUTs with Non-Abelian Flavor Symmetries

In this section, we want to extend our choice of Abelian flavor symmetries G_A to non-Abelian ones $G_f = G_A \ltimes G_B$, where G_A are the flavor symmetries shown in Fig. 8.1 and $G_B = G_{B_1} \times G_{B_2} \times G_{B_3} \times ...$ can also be products of cyclic groups. For this upgraded model setup, we want to give a geometrical interpretation in 5D. However, note that the 5D setup serves "just" as a interpretation of flavor symmetry breaking, it is not a necessary ingredient for the models. The geometry consists of two 5D throats in the flat limit [119, 120] of length πR_1 and πR_2, with $1/R_{1,2} \gtrsim M_{\text{GUT}} \simeq 10^{16}$ GeV, which are glued together at a common point, the ultraviolet (UV) brane. The endpoints at $y_1 = \pi R_1$ and $y_2 = \pi R_2$ are called infrared (IR) branes. This setup is shown in Fig. 8.2. The matter field zero modes $\mathbf{10}_i, \overline{\mathbf{5}}_i$, and $\mathbf{1}_i$, are symmetrically localized at

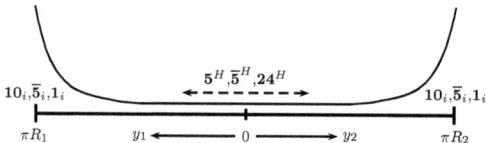

Figure 8.2: 5D SUSY $SU(5)$ GUTs on two 5D throats. The matter field zero modes $\mathbf{10}_i, \overline{\mathbf{5}}_i$, and $\mathbf{1}_i$, are symmetrically located at $y_1 = \pi R_1$ and $y_2 = \pi R_2$. The Higgs hypermultiplets $\mathbf{5}^H, \overline{\mathbf{5}}^H, \mathbf{24}^H$, and the gauge supermultiplet, can propagate freely within the two throat setup.

$y_1 = \pi R_1$ and $y_2 = \pi R_2$ by introducing suitable bulk fermion masses [119, 121] and boundary conditions. This allows a description of the fermion superfield zero modes in usual 4D $N = 1$ SUSY assuming that some 4D $N = 2$ SUSY (which is equivalent to minimal 5D SUSY) is locally broken to 4D $N = 1$ SUSY at the UV/IR branes. Contrary to this localization, the $SU(5)$ gauge supermultiplet and the Higgs hypermultiplets $\mathbf{5}^H$ and $\overline{\mathbf{5}}^H$ are freely propagating in the two throats. Thereby, the $\mathbf{24}^H$ bulk Higgs hypermultiplet acquires a VEV in the hypercharge direction $\langle \mathbf{24}^H \rangle \propto \text{diag}(-\frac{1}{2}, -\frac{1}{2}, \frac{1}{3}, \frac{1}{3}, \frac{1}{3})$. This breaks[1] $SU(5)$ down to the SM gauge group and we obtain an explanation for the SM quantum numbers and charge quantization. Summarizing this, we can introduce the 5D Lagrangian for the SUSY $SU(5)$ Yukawa couplings of the zero mode fermions as

$$\mathcal{L}_{5D}^{SU(5)} = \int d^2\theta \Big[\delta(y_1 - \pi R_1)\Big(\tilde{Y}_{ij,R_1}^u \mathbf{10}_i \mathbf{10}_j \mathbf{5}^H + \tilde{Y}_{ij,R_1}^d \mathbf{10}_i \overline{\mathbf{5}}_j \overline{\mathbf{5}}^H + \tilde{Y}_{ij,R_1}^D \overline{\mathbf{5}}_i \mathbf{1}_j \mathbf{5}^H \\
+ M_R \tilde{Y}_{ij,R_1}^R \mathbf{1}_i \mathbf{1}_j \Big) + \delta(y_2 - \pi R_2)\Big(\tilde{Y}_{ij,R_2}^u \mathbf{10}_i \mathbf{10}_j \mathbf{5}^H + \tilde{Y}_{ij,R_2}^d \mathbf{10}_i \overline{\mathbf{5}}_j \overline{\mathbf{5}}^H \\
+ \tilde{Y}_{ij,R_2}^D \overline{\mathbf{5}}_i \mathbf{1}_j \mathbf{5}^H + M_R \tilde{Y}_{ij,R_2}^R \mathbf{1}_i \mathbf{1}_j \Big) + \text{H.c.} \Big], \quad (8.11)$$

where $\tilde{Y}_{ij,R_1}^x$ and $\tilde{Y}_{ij,R_2}^x$ ($x = u, d, D, R$) are the Yukawa matrices (with mass dimension $-1/2$) and $M_R \simeq 10^{14}$ GeV is the $B - L$ breaking scale. This leads to the dimensionless low energy Yukawa coupling matrices $Y_{ij}^x = (M_* \pi R)^{-1/2} (\tilde{Y}_{ij,R_1}^x + \tilde{Y}_{ij,R_2}^x)$ of Eq. (8.11), where $M_* \simeq (M_{\text{Pl}}^2 R_{1,2}^{-1})^{1/3}$ and M_{Pl} is the usual 4D Planck mass.

Now that we have fixed the setup, we want to apply this to our example of Sec. 8.1. Note that the top Yukawa coupling is unsuppressed by the flavor symmetry and can be large without

---
[1]In usual 5D GUT models, $SU(5)$ is broken by boundary conditions [122].

8.3 5D SUSY $SU(5)$ GUTs with Non-Abelian Flavor Symmetries

requiring strong coupling as long as $M_* R_{1,2} \lesssim 16\pi^2$ [123]. The flavor symmetry in this model is $G_A = Z_3 \times Z_8 \times Z_9$. This flavor symmetry will now be extended to

$$G_f = G_A \ltimes G_B = G_A \ltimes (G_{B_1} \times G_{B_2} \times G_{B_3}) , \qquad (8.12)$$

where G_{B_i} are supposed to be the following discrete symmetries

$$G_{B_1} : \bar{5}_2 \leftrightarrow \bar{5}_3, \quad G_{B_2} : 1_1 \leftrightarrow 1_3, \quad G_{B_3} : 10_3 \rightarrow -10_3 . \qquad (8.13)$$

The result is that G_A controls the order of magnitude of the mass matrices, i.e., it leads to the textures presented in Sec. 8.1, while G_B induces for the Yukawa couplings the exact relations

$$Y_{32}^d = Y_{33}^d, \quad Y_{21}^D = Y_{23}^D, \quad Y_{31}^D = Y_{33}^D, \quad Y_{11}^R = Y_{33}^R . \qquad (8.14)$$

G_A is spontaneously broken by singly charged flavon fields located at the IR branes (for a 4+3 dimensional model with flavor symmetry breaking by Wilson lines see Ref. [124]). The breaking of G_B in order to obtain the right predictions for the Yukawa couplings can be done at both ends of the throats since the 4D Yukawa couplings receive contributions from both, see Eq. (8.11). Therefore, we can break G_B at $y_1 = \pi R_1$ to G_{B_1} and at $y_2 = \pi R_2$ to $G_{B_2} \times G_{B_3}$. This symmetry breaking is illustrated in Fig. 8.3.

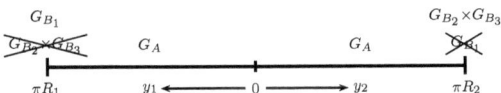

Figure 8.3: Local breaking of the flavor symmetry $G_f = G_A \ltimes (G_{B_1} \times G_{B_2} \times G_{B_3})$ to G_A at the IR branes.

In Sec. 8.1, we have already presented the resulting PMNS mixing angles for $G_f = G_A$. However, G_A predicts only the texture, while $G_f = G_A \ltimes G_B$ also controls the Yukawa couplings. This allows us to see the effects of the transition from Abelian flavor symmetries to non-Abelian ones. For this, we have plotted the distribution of θ_{23} in Fig. 8.4 for a random variation of all Yukawa couplings in Eqs. (8.9a) and (8.9b) up to 10% with $G_f = G_A$ (left) and $G_f = G_A \ltimes G_B$ (right). In Fig. 8.4, we can see the effect of the non-Abelian nature of the flavor symmetry on θ_{23}. The leading order term of the sum rule for the atmospheric mixing angle, i.e., $\theta_{23} \approx \pi/4 + \epsilon/\sqrt{2}$ is exactly predicted by the flavor symmetry. In addition, we obtain for the first time the QLC

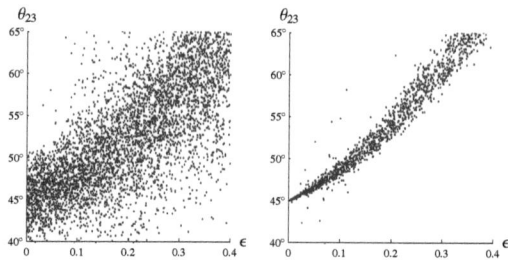

Figure 8.4: Effect of the non-Abelian flavor symmetry $G_A \ltimes G_B$ (right) on θ_{23} compared to the Abelian flavor symmetry G_A (left) for a 10% variation of all Yukawa couplings. The right plot illustrates the exact prediction of the zeroth order term $\pi/4$ in the sum rule $\theta_{23} = \pi/4 + \epsilon/\sqrt{2}$ due to the non-Abelian nature of the flavor symmetry.

relation $\theta_{12} \approx \pi/4 - \epsilon/\sqrt{2}$ in an $SU(5)$ GUT.

Since $SU(5)$ is broken by a bulk Higgs field, the bosons of the broken gauge symmetries acquire GUT scale masses and consequently, all fermion zero modes can be localized at the IR branes without introducing rapid proton decay through $d = 6$ operators. Doublet-triplet splitting and suppressing $d = 5$ proton decay may, e.g., be achieved by resorting to suitable extensions of the Higgs sector [125, 126]. Again, since our flavor symmetry is global, quantum gravity effects might require it to be gauged [111]. Anomalies can then be canceled via Chern-Simons terms in the 5D bulk.

8.4 Summary

In this chapter, we have extended our approach of Chap. 6 in various ways. First, we have considered additionally the quark sector within a $SU(5)$ GUT scenario and added to each seesaw representation the corresponding quark mass matrices, which we have used as our new reference list. Second, we have followed the method of Chap. 6 and generated systematically all flavor charge combinations for numerous flavor symmetries that are direct products of cyclic groups. As a result we obtained 437 $SU(5)$ GUTs with Abelian flavor symmetries [112]. Each model yields an excellent fit to fermion masses, quark mixing and nearly tribimaximal lepton mixing. We have presented an explicit example how the Abelian flavor symmetries can be extended to non-Abelian groups and an embedding of the model into a 5D scenario, which allows a geometrical interpretation of the flavor symmetry breaking. For this example, we have

8.4 Summary

shown the impact of the non-Abelian nature of the flavor symmetry on the atmospheric mixing angle as compared to the Abelian group. In addition, we have obtained, for the first time, the QLC relation $\theta_{12} \approx \pi/4 - \epsilon/\sqrt{2}$ in an explicit $SU(5)$ GUT.

Chapter 9

Summary and Outlook

In physics, symmetries are a fundamental concept to explain the properties and behavior of particles. This is formulated in the SM of particle physics. But it leaves still some open questions such as why do we have three generations of particles with such a strong hierarchy in the quark sector, why are the quark and lepton mixings so different or what is the structure of the Higgs sector. In order to address these open questions of the SM and to predict future experimental results, different approaches are considered. One possibility are GUTS such as $SU(5)$ or $SO(10)$. GUTs are vertical symmetries since they accommodate the SM particles into multiplets. A convincing, complementary way is to introduce a horizontal symmetry, *i.e.*, a flavor symmetry which can serve as an explanation for the strong mass hierarchy among quarks and charged leptons, the different mixings in the quark and lepton sectors. In combination with the seesaw mechanism, predicted by some GUTs, also the smallness of neutrino masses can be understood and physics near the GUT scale can be probed by neutrinos. In this book, we have combined of all three elements, *i.e.*, GUTs, seesaw mechanism, and flavor symmetries.

An appealing approach is to use discrete non-Abelian flavor symmetries such as A_4 or T' since they are very predictive. However, in general, such models especially in combination with GUTs, usually have a very complicated Higgs sector and need fine-tuning. An alternative is provided by the Froggatt-Nielsen mechanism having a comparatively simple scalar sector. In addition, discrete flavor symmetries such as $\Delta(27)$, A_4,... are isomorphic to semi-direct products of cyclic groups. For example, A_4 is isomorphic to the semi-direct product $Z_3 \ltimes (Z_2 \times$

Z_2). Therefore, we have used the FN mechanism for products of cyclic groups to construct lepton flavor models. All of these lepton models lead to nearly tribimaximal mixing, have a simple scalar sector, and use the type-I seesaw without fine-tuning. The so called tribimaximal mixing scenario, which predicts $\sin^2 \theta_{12} = 1/3$, zero U_{e3} and maximal θ_{23}, perfectly fit current experimental data. Small breaking terms in the mass matrices generates deviations from the tribimaximal scheme and lead to testable correlations between the parameters.

All models exhibit our hypothesis of extended quark-lepton complementarity stating that all masses and mixing are some powers of the Cabibbo angle, where the zeroth order for the mixings is interpreted as maximal mixing. This is motivated by the observed quark mixing matrix as well as μ-τ-symmetry. The models include special cases, often discussed in literature, such as diagonal charged lepton and right-handed neutrino mass matrices, respectively. But we also obtain models with maximal mixing among charged leptons as well as neutrinos, where the mixing in the neutrino sector can stem form both, the Dirac and the Majorana neutrino mass matrix. In addition, this general approach reveals new features such as new mass matrix textures and new sum rules.

In an approach similar to the one described above, we have constructed SUSY $SU(5)$ GUTs with products of cyclic flavor symmetries. All of these GUT models predict the CKM matrix and not the unit matrix as often stated in literature, a nearly tribimaximal PMNS matrix as well as all quark and lepton mass hierarchies. Also the extension to non-Abelian semi-direct products of Z_n symmetries was done. Thereby, we could show the effect of non-Abelian flavor symmetries on the mixing angles as compared to the Abelian case. In order to obtain a geometrical interpretation of the symmetry breaking, the model has been extended to a simple 5D SUSY $SU(5)$ GUT on two throats.

The lepton models and the SUSY $SU(5)$ GUTs are based on the type-I seesaw mechanism. Therefore, interesting phenomenological implications arise in addition to neutrino oscillation experiments and $0\nu\beta\beta$ experiments. Lepton flavor violating processes for example. These were investigated for the LHC relevant benchmark scenario SPS1a' in mSUGRA. For this purpose, we have complexified the textures by introducing random CP phases at the Lagrangian level and analyzed the resulting LFV decay rates. This leads to an enhancement of the LFV rates of typically an order of magnitude as compared to the CP conserving case. Moreover, the branching ratios exhibit a strong dependence on the choice of the phases in the Lagrangian, i.e., the LFV rates appear to be essentially uncorrelated with the possible high- and low energy lepton mixing parameters.

Apart from the work done so far, further phenomenological investigations would be interesting in order to select the most promising models and to extend the existing framework. One phenomenological possibility is to include leptogenesis. Also from the model building perspective, different approaches could be considered. One possibility would be to consider embedding of discrete flavor symmetries into continuous ones and gauge them in order to avoid potentially harmful quantum gravity effects. This could also be combined with GUT scenarios which are more predictive than $SU(5)$. Another approach could arise from the group theoretical side. There, a more detailed analysis of flavor symmetries and their isomorphisms would bring a greater insight into (systematic) model building. These three direction, phenomenology, GUTs, and group theory, are only a selection. Also interesting are other topics such as a more detailed analysis of our obtained textures, model building implications of flavor symmetries and for our models in particular, astrophysical and LHC/ILC phenomenology as well as adjoined topics, where we can gain new insights.

Appendix A

Appendix

A.1 Supplementary Information for Seesaw Realizations

Here, we give supplementary information to Table 5.1 in order to be able to reproduce the mass matrix textures and realizations. These contain, besides the phases of U_ℓ, U_D, $U_{D'}$, and U_R, the PMNS mixing angles θ_{ij}, the performance indicator χ^2 defined in Eq. (5.4), and in the last column the number of realizations leading to this specific texture set. Note that we choose $\varphi^{D'}_{1,2,3} = \alpha^D_{1,2} = 0$, since these phases appear in $M^{\text{th}}_{\text{eff}}$ only in combination with other phases and can thus be absorbed (see Sec. 4.2.1).

#	$(\delta^l, \alpha^l_1, \alpha^l_2)$	$(\delta^D, \varphi^D_1, \varphi^D_2, \varphi^D_3)$	$(\delta^{D'}, \alpha^{D'}_1, \alpha^{D'}_2)$	$(\delta^R, \varphi^R_1, \varphi^R_2, \varphi^R_3)$	$(\theta_{12}, \theta_{13}, \theta_{23})$	χ^2	Cases
1	$(0, \pi, 0)$	$(\pi, 0, 0, \pi)$	(π, π, π)	$(0, 0, 0, \pi)$	$(34.0°, 0.2°, 52.2°)$	7.12	18
2	$(\pi, 0, \pi)$	$(\pi, 0, 0, \pi)$	$(0, 0, \pi)$	$(\pi, 0, \pi, \pi)$	$(33.6°, 0.2°, 51.5°)$	5.29	38
3	$(\pi, 0, 0)$	$(0, 0, \pi, 0)$	$(0, 0, 0)$	$(\pi, 0, \pi, 0)$	$(33.5°, 0.2°, 51.3°)$	4.9	26
4	$(0, 0, \pi)$	$(0, 0, 0, \pi)$	$(\pi, 0, \pi)$	$(\pi, 0, \pi, \pi)$	$(33.5°, 0.1°, 51.2°)$	4.71	17
5	$(0, 0, 0)$	$(0, 0, 0, 0)$	$(0, \pi, 0)$	$(0, 0, 0, \pi)$	$(33.0°, 0.4°, 51.2°)$	4.7	17
6	$(\pi, \pi, 0)$	$(\pi, 0, 0, 0)$	(π, π, π)	$(0, 0, 0, 0)$	$(33.3°, 0.4°, 51.2°)$	4.7	177
7	$(\pi, \pi, 0)$	$(\pi, 0, 0, 0)$	$(0, 0, 0)$	$(\pi, 0, 0, 0)$	$(33.3°, 0.4°, 51.2°)$	4.7	63
8	$(0, 0, \pi)$	$(0, 0, 0, \pi)$	$(0, \pi, 0)$	$(0, 0, \pi, \pi)$	$(33.5°, 0.1°, 51.2°)$	4.71	17
9	(π, π, π)	$(0, 0, \pi, 0)$	$(\pi, 0, 0)$	$(0, 0, \pi, 0)$	$(32.9°, 0.2°, 51.2°)$	4.76	23

10	$(\pi,0,0)$	$(\pi,0,0,\pi)$	$(0,0,0)$	$(0,0,0,\pi)$	$(33.2°, 0.2°, 51.3°)$	4.78	597
11	$(\pi,0,0)$	$(\pi,0,0,\pi)$	$(0,0,0)$	$(0,0,0,\pi)$	$(33.2°, 0.2°, 51.3°)$	4.78	835
12	$(0,\pi,\pi)$	$(0,0,0,0)$	$(0,\pi,0)$	$(\pi,0,0,\pi)$	$(33.4°, 0.0°, 51.3°)$	4.81	475
13	$(0,0,0)$	$(\pi,0,0,\pi)$	$(0,0,0)$	$(\pi,0,0,0)$	$(33.4°, 0.2°, 51.3°)$	4.84	104
14	$(\pi,0,0)$	$(\pi,0,\pi,0)$	$(\pi,\pi,0)$	$(0,0,0,\pi)$	$(33.5°, 0.6°, 51.3°)$	4.89	14
15	$(\pi,0,0)$	$(0,0,\pi,0)$	$(0,0,0)$	$(\pi,0,\pi,0)$	$(33.5°, 0.2°, 51.3°)$	4.9	17
16	$(\pi,0,0)$	$(\pi,0,\pi,0)$	$(0,\pi,0)$	$(0,0,0,\pi)$	$(33.5°, 0.6°, 51.3°)$	4.9	120
17	$(\pi,0,0)$	$(\pi,0,0,0)$	$(0,0,\pi)$	$(\pi,0,0,\pi)$	$(33.3°, 0.0°, 51.4°)$	4.96	1138
18	$(0,\pi,\pi)$	$(0,0,\pi,\pi)$	$(0,\pi,\pi)$	$(0,0,\pi,\pi)$	$(33.5°, 0.2°, 51.4°)$	4.97	9
19	$(0,\pi,\pi)$	$(0,0,0,0)$	$(0,0,0)$	$(0,0,0,0)$	$(33.4°, 0.0°, 51.4°)$	4.99	387
20	$(\pi,0,\pi)$	$(\pi,0,0,0)$	$(0,\pi,0)$	$(0,0,\pi,\pi)$	$(33.7°, 0.2°, 51.2°)$	5.02	26
21	$(\pi,0,0)$	$(0,0,0,\pi)$	$(0,0,0)$	$(0,0,0,\pi)$	$(33.4°, 0.1°, 51.5°)$	5.03	776
22	$(\pi,0,0)$	$(0,0,0,\pi)$	$(0,0,\pi)$	$(\pi,0,\pi,\pi)$	$(33.2°, 0.1°, 51.5°)$	5.03	876
23	$(\pi,0,0)$	$(0,0,0,\pi)$	$(\pi,\pi,0)$	$(0,0,\pi,\pi)$	$(33.2°, 0.1°, 51.5°)$	5.03	1351
24	(π,π,π)	$(0,0,\pi,0)$	$(0,\pi,\pi)$	$(\pi,0,\pi,\pi)$	$(33.5°, 0.3°, 51.4°)$	5.05	27
25	$(\pi,0,0)$	$(0,0,\pi,0)$	$(\pi,\pi,0)$	$(0,0,0,\pi)$	$(33.5°, 0.1°, 51.4°)$	5.06	392
26	$(\pi,0,0)$	$(0,0,0,\pi)$	$(0,0,0)$	$(0,0,0,\pi)$	$(33.4°, 0.1°, 51.5°)$	5.09	307
27	$(0,0,\pi)$	$(0,0,0,\pi)$	$(0,\pi,0)$	$(\pi,0,\pi,0)$	$(33.5°, 0.2°, 51.4°)$	5.09	26
28	$(\pi,0,0)$	$(0,0,\pi,0)$	$(\pi,0,0)$	$(\pi,0,\pi,0)$	$(33.7°, 0.2°, 51.3°)$	5.11	296
29	$(\pi,0,\pi)$	$(\pi,0,0,\pi)$	(π,π,π)	$(0,0,0,\pi)$	$(33.6°, 0.2°, 51.4°)$	5.16	5
30	$(\pi,0,\pi)$	$(\pi,0,0,\pi)$	(π,π,π)	$(0,0,0,\pi)$	$(33.6°, 0.2°, 51.4°)$	5.16	5
31	$(\pi,0,\pi)$	$(\pi,0,0,\pi)$	$(0,0,\pi)$	$(\pi,0,\pi,\pi)$	$(33.6°, 0.2°, 51.5°)$	5.29	38
32	$(\pi,0,0)$	$(0,0,\pi,0)$	$(0,\pi,\pi)$	$(\pi,0,0,\pi)$	$(33.3°, 0.2°, 51.7°)$	5.31	343
33	(π,π,π)	$(\pi,0,0,0)$	$(0,0,0)$	$(0,0,\pi,0)$	$(33.6°, 0.1°, 51.5°)$	5.31	83
34	(π,π,π)	$(\pi,0,0,0)$	$(0,0,0)$	$(0,0,\pi,0)$	$(33.6°, 0.1°, 51.5°)$	5.31	81
35	$(0,\pi,0)$	$(\pi,0,0,\pi)$	$(0,\pi,0)$	$(\pi,0,\pi,0)$	$(33.7°, 0.1°, 51.5°)$	5.32	143
36	$(\pi,0,\pi)$	$(\pi,0,0,\pi)$	$(\pi,0,0)$	$(0,0,0,\pi)$	$(32.9°, 0.2°, 51.6°)$	5.33	17
37	$(\pi,0,\pi)$	$(\pi,0,0,\pi)$	$(0,\pi,\pi)$	$(0,0,0,0)$	$(32.9°, 0.2°, 51.6°)$	5.33	17
38	$(\pi,0,0)$	$(0,0,0,\pi)$	$(\pi,0,0)$	$(0,0,0,\pi)$	$(33.2°, 0.1°, 51.7°)$	5.33	17
39	$(\pi,0,0)$	$(0,0,0,\pi)$	$(\pi,0,0)$	$(0,0,0,\pi)$	$(33.2°, 0.1°, 51.7°)$	5.33	33
40	$(\pi,0,0)$	$(0,0,0,0)$	$(0,0,\pi)$	$(\pi,0,0,\pi)$	$(33.2°, 0.1°, 51.7°)$	5.37	17
41	$(\pi,0,\pi)$	$(\pi,0,0,\pi)$	$(\pi,0,0)$	$(0,0,\pi,\pi)$	$(33.1°, 0.0°, 51.8°)$	5.47	26

A.1 Supplementary Information for Seesaw Realizations

42	$(\pi,0,\pi)$	$(\pi,0,0,\pi)$	$(0,\pi,\pi)$	$(0,0,0,\pi)$	$(33.1°, 0.0°, 51.8°)$	5.47	17
43	$(0,0,\pi)$	$(\pi,0,\pi,0)$	$(\pi,0,0)$	$(0,0,0,\pi)$	$(33.3°, 0.2°, 51.8°)$	5.48	5
44	$(0,0,\pi)$	$(\pi,0,\pi,0)$	$(\pi,0,0)$	$(0,0,0,\pi)$	$(33.3°, 0.2°, 51.8°)$	5.48	14
45	$(\pi,0,\pi)$	$(\pi,0,0,\pi)$	$(0,\pi,\pi)$	$(\pi,0,\pi,\pi)$	$(34.0°, 0.3°, 51.3°)$	5.66	18
46	$(0,0,0)$	$(\pi,0,0,\pi)$	$(0,0,0)$	$(\pi,0,0,\pi)$	$(33.9°, 0.7°, 51.5°)$	5.84	5
47	$(0,0,0)$	$(\pi,0,0,\pi)$	$(\pi,0,0)$	$(0,0,0,0)$	$(34.0°, 0.7°, 51.5°)$	5.96	5
48	(π,π,π)	$(0,0,0,\pi)$	$(0,\pi,\pi)$	$(0,0,0,0)$	$(33.7°, 0.1°, 51.9°)$	5.98	9
49	$(\pi,0,\pi)$	$(\pi,0,0,0)$	(π,π,π)	$(0,0,0,\pi)$	$(33.5°, 0.4°, 52.0°)$	6.02	34
50	$(\pi,0,\pi)$	$(\pi,0,0,0)$	$(0,0,0)$	$(\pi,0,\pi,0)$	$(33.5°, 0.4°, 52.0°)$	6.02	26
51	$(\pi,0,\pi)$	$(0,0,0,0)$	$(\pi,0,0)$	$(0,0,0,\pi)$	$(33.8°, 0.3°, 51.9°)$	6.19	86
52	$(\pi,0,\pi)$	$(0,0,0,0)$	$(\pi,0,0)$	$(0,0,0,\pi)$	$(33.8°, 0.3°, 51.9°)$	6.19	87
53	$(\pi,0,\pi)$	$(0,0,0,0)$	$(0,\pi,\pi)$	$(0,0,\pi,\pi)$	$(33.8°, 0.3°, 51.9°)$	6.19	108
54	$(0,0,0)$	$(0,0,\pi,0)$	$(0,\pi,\pi)$	$(0,0,\pi,\pi)$	$(34.4°, 0.1°, 51.1°)$	6.26	14
55	(π,π,π)	$(\pi,0,\pi,\pi)$	$(0,0,0)$	$(0,0,\pi,\pi)$	$(34.4°, 0.1°, 51.1°)$	6.26	14
56	$(0,0,0)$	$(\pi,0,0,\pi)$	$(\pi,\pi,0)$	$(\pi,0,0,\pi)$	$(32.5°, 0.2°, 52.0°)$	6.44	79
57	$(0,0,0)$	$(\pi,0,0,\pi)$	$(\pi,\pi,0)$	$(\pi,0,0,\pi)$	$(32.5°, 0.2°, 52.0°)$	6.44	34
58	$(0,\pi,\pi)$	$(\pi,0,0,\pi)$	$(0,0,0)$	$(0,0,\pi,0)$	$(34.2°, 0.7°, 51.5°)$	6.46	5
59	$(0,\pi,0)$	$(0,0,\pi,0)$	$(0,0,\pi)$	$(0,0,0,\pi)$	$(34.0°, 0.2°, 52.0°)$	6.59	5
60	$(0,0,0)$	$(\pi,0,0,\pi)$	$(\pi,0,\pi)$	$(0,0,0,\pi)$	$(34.0°, 0.4°, 52.0°)$	6.77	5
61	$(0,\pi,0)$	$(\pi,0,0,\pi)$	$(0,0,0)$	$(0,0,\pi,0)$	$(34.0°, 0.2°, 52.2°)$	7.12	18
62	$(0,\pi,0)$	$(\pi,0,0,\pi)$	(π,π,π)	$(0,0,0,\pi)$	$(34.0°, 0.2°, 52.2°)$	7.12	18
63	$(0,\pi,0)$	$(\pi,0,0,\pi)$	$(0,0,0)$	$(0,0,\pi,0)$	$(34.0°, 0.2°, 52.2°)$	7.12	18
64	(π,π,π)	$(\pi,0,0,0)$	$(0,0,\pi)$	$(0,0,\pi,\pi)$	$(34.1°, 0.2°, 52.2°)$	7.25	17
65	$(\pi,0,\pi)$	$(0,0,\pi,\pi)$	$(0,0,\pi)$	$(0,0,0,0)$	$(34.5°, 0.3°, 51.8°)$	7.84	9
66	$(0,\pi,0)$	$(\pi,0,\pi,\pi)$	$(0,\pi,0)$	$(0,0,0,0)$	$(34.5°, 0.3°, 51.8°)$	7.84	9
67	(π,π,π)	$(0,0,0,0)$	$(0,\pi,\pi)$	$(0,0,\pi,\pi)$	$(34.9°, 0.4°, 51.8°)$	9.31	26
68	(π,π,π)	$(0,0,0,0)$	$(\pi,0,0)$	$(0,0,0,0)$	$(34.9°, 0.4°, 51.8°)$	9.31	31
69	$(0,0,0)$	$(\pi,0,0,0)$	$(0,0,0)$	$(0,0,\pi,\pi)$	$(34.9°, 0.4°, 51.8°)$	9.31	26
70	$(0,0,0)$	$(\pi,0,0,0)$	(π,π,π)	$(0,0,0,0)$	$(34.9°, 0.4°, 51.8°)$	9.31	31
71	$(0,0,0)$	$(\pi,0,0,0)$	$(0,\pi,0)$	$(0,0,\pi,0)$	$(35.3°, 0.3°, 51.3°)$	10.73	17
72	$(0,0,0)$	$(\pi,0,0,0)$	$(0,\pi,0)$	$(0,0,\pi,0)$	$(35.3°, 0.3°, 51.3°)$	10.73	17

Table A.1: Supplementary information for seesaw realizations of Table 5.1 ($\varphi^{D'}_{1,2,3} = \alpha^D_{1,2} = 0$).

A.2 Supplementary Information for Lepton Flavor Models

Here, we give supplementary information to the models in Table 6.1. By using the data in Table A.2, a full reconstruction of the mass matrices of the 22 models following the notation of Sec. 6.1 is possible (for further detailed examples on such reconstructions, see also Ref. [75]). Note that we choose $\varphi^\ell_i = \varphi^{\ell'}_i = \varphi^{D'}_i = \alpha^D_j = 0$ for $i = 1, 2, 3$ and $j = 1, 2$, since these phases appear in $M^{\text{th}}_{\text{eff}}$ only in combination with other phases and can thus be absorbed (see Sec. 4.2.1).

#	m^D_i/m_D m^R_i/M_{B-L}	$(\theta^\ell_{12}, \theta^\ell_{13}, \theta^\ell_{23})$ $(\delta^\ell, \alpha^\ell_1, \alpha^\ell_2)$	$(\theta^{\ell'}_{12}, \theta^{\ell'}_{13}, \theta^{\ell'}_{23})$ $(\delta^{\ell'}, \alpha^{\ell'}_1, \alpha^{\ell'}_2)$	$(\theta^D_{12}, \theta^D_{13}, \theta^D_{23})$ $(\delta^D, \varphi^D_1, \varphi^D_2, \varphi^D_3)$	$(\theta^{D'}_{12}, \theta^{D'}_{13}, \theta^{D'}_{23})$ $(\delta^{D'}, \alpha^{D'}_1, \alpha^{D'}_2)$	$(\theta^R_{12}, \theta^R_{13}, \theta^R_{23})$ $(\delta^R, \varphi^R_1, \varphi^R_2, \varphi^R_3)$
1	$(\epsilon, 1, \epsilon)$ $(\epsilon, 1, 1)$	$(\epsilon^2, \epsilon^2, \epsilon^2)$ (π, π, π)	$(\frac{\pi}{4}, \epsilon^2, 0)$ $(0, 0, 0)$	$(0, \epsilon^2, \frac{\pi}{4})$ $(0, 0, 0, 0)$	$(\epsilon, \epsilon, \epsilon^2)$ $(0, 0, 0)$	$(\epsilon^2, \frac{\pi}{4}, \epsilon^2)$ $(\pi, 0, \pi, 0)$
2	$(\epsilon, 1, \epsilon)$ $(\epsilon, 1, 1)$	$(\epsilon^2, \epsilon^2, \frac{\pi}{4})$ $(0, 0, 0)$	$(\epsilon, 0, \epsilon^2)$ $(0, 0, 0)$	$(\epsilon^2, \frac{\pi}{4}, \epsilon^2)$ $(0, 0, 0, 0)$	$(\epsilon, \epsilon^2, 0)$ $(0, \pi, \pi)$	$(\epsilon, 0, \frac{\pi}{4})$ $(0, 0, 0, 0)$
3	$(\epsilon, 1, \epsilon)$ $(\epsilon, 1, 1)$	$(\epsilon, 0, \epsilon^2)$ $(0, 0, 0)$	$(\epsilon, \epsilon, \epsilon^2)$ $(0, 0, 0)$	$(\epsilon, \frac{\pi}{4}, \frac{\pi}{4})$ $(\pi, 0, 0, \pi)$	$(\epsilon, \epsilon, \frac{\pi}{4})$ $(0, \pi, \pi)$	$(\epsilon, \epsilon, \frac{\pi}{4})$ $(0, 0, 0, \pi)$
4	$(\epsilon, \epsilon, 1)$ $(\epsilon, 1, 1)$	$(\epsilon, \epsilon, \epsilon^2)$ $(0, \pi, \pi)$	$(\frac{\pi}{4}, \epsilon^2, 0)$ $(0, 0, 0)$	$(\frac{\pi}{4}, \epsilon^2, \frac{\pi}{4})$ $(0, 0, \pi, 0)$	$(0, \epsilon^2, \frac{\pi}{4})$ $(0, 0, 0)$	$(0, \epsilon, \epsilon^2)$ $(0, 0, \pi, 0)$
5	$(\epsilon, 1, \epsilon)$ $(\epsilon, 1, 1)$	$(0, \epsilon, \epsilon^2)$ $(0, \pi, 0)$	$(\epsilon, 0, \epsilon)$ $(0, 0, 0)$	$(\epsilon, \frac{\pi}{4}, \frac{\pi}{4})$ $(\pi, 0, 0, \pi)$	$(\epsilon^2, \frac{\pi}{4}, \epsilon^2)$ $(0, \pi, \pi)$	$(0, \frac{\pi}{4}, 0)$ $(0, 0, 0, \pi)$
6	$(\epsilon^2, 1, \epsilon)$ $(\epsilon^2, \epsilon, 1)$	$(0, \epsilon, \epsilon^2)$ $(0, \pi, 0)$	$(\epsilon, 0, \epsilon)$ $(0, 0, 0)$	$(\epsilon, \frac{\pi}{4}, \frac{\pi}{4})$ $(\pi, 0, 0, \pi)$	$(\epsilon, \epsilon, \frac{\pi}{4})$ $(\pi, 0, 0)$	$(\epsilon, \epsilon, \frac{\pi}{4})$ $(\pi, 0, \pi, 0)$
7	$(\epsilon^2, \epsilon, 1)$ $(\epsilon^2, \epsilon, 1)$	$(0, \epsilon^2, \frac{\pi}{4})$ $(0, \pi, \pi)$	$(\epsilon, 0, \epsilon^2)$ $(0, 0, \pi)$	$(\frac{\pi}{4}, \epsilon^2, \epsilon)$ $(0, 0, 0, \pi)$	$(0, \frac{\pi}{4}, \epsilon)$ $(0, 0, \pi)$	$(\epsilon^2, \frac{\pi}{4}, \epsilon^2)$ $(\pi, 0, \pi, \pi)$
8	$(\epsilon^2, 1, \epsilon)$ $(\epsilon^2, \epsilon, 1)$	$(\epsilon^2, \epsilon, \epsilon^2)$ $(0, \pi, 0)$	$(\epsilon, \epsilon^2, 0)$ $(0, 0, 0)$	$(\epsilon, \frac{\pi}{4}, \frac{\pi}{4})$ $(\pi, 0, 0, \pi)$	$(0, \epsilon^2, \frac{\pi}{4})$ $(0, 0, \pi)$	$(\epsilon^2, \epsilon^2, \frac{\pi}{4})$ $(\pi, 0, \pi, 0)$
9	$(\epsilon, 1, \epsilon)$ $(\epsilon, 1, 1)$	$(\epsilon, 0, \epsilon^2)$ $(0, 0, 0)$	$(\epsilon, \epsilon, 0)$ $(0, 0, 0)$	$(\epsilon, \frac{\pi}{4}, \frac{\pi}{4})$ $(\pi, 0, 0, \pi)$	$(\frac{\pi}{4}, \epsilon, \epsilon^2)$ $(\pi, 0, \pi)$	$(\frac{\pi}{4}, \epsilon, 0)$ $(0, 0, 0, \pi)$

A.2 Supplementary Information for Lepton Flavor Models

10	$(\epsilon^2,\epsilon,1)$	$(\epsilon^2,\epsilon^2,\epsilon^2)$	$(\frac{\pi}{4},\epsilon^2,0)$	$(\frac{\pi}{4},\epsilon^2,\frac{\pi}{4})$	$(\epsilon^2,0,\epsilon^2)$	$(\epsilon,\epsilon^2,0)$
	$(\epsilon^2,\epsilon,1)$	(π,π,π)	$(0,0,0)$	$(0,0,0,\pi)$	$(0,0,0)$	$(0,0,0,0)$
11	$(\epsilon,1,\epsilon)$	$(\epsilon,\epsilon^2,\epsilon^2)$	$(\frac{\pi}{4},0,\epsilon^2)$	$(\epsilon,\frac{\pi}{4},\frac{\pi}{4})$	$(\epsilon,\epsilon^2,\epsilon)$	$(\epsilon,\epsilon^2,\epsilon)$
	$(\epsilon,1,1)$	$(\pi,0,0)$	$(0,0,0)$	$(\pi,0,0,\pi)$	$(\pi,0,0)$	$(\pi,0,0,0)$
12	$(\epsilon,1,\epsilon)$	$(0,\epsilon,\epsilon^2)$	$(\epsilon,0,0)$	$(\epsilon,\frac{\pi}{4},\frac{\pi}{4})$	$(\epsilon^2,\frac{\pi}{4},\epsilon)$	$(\epsilon,\frac{\pi}{4},\epsilon^2)$
	$(\epsilon,1,1)$	$(0,\pi,0)$	$(0,0,0)$	$(\pi,0,0,\pi)$	$(0,\pi,0)$	$(\pi,0,\pi,0)$
13	$(\epsilon^2,\epsilon,1)$	$(\epsilon^2,\epsilon^2,\epsilon^2)$	$(\epsilon,0,0)$	$(\frac{\pi}{4},\epsilon^2,\frac{\pi}{4})$	$(\epsilon,\epsilon^2,0)$	$(0,\epsilon^2,\epsilon^2)$
	$(\epsilon^2,\epsilon,1)$	$(\pi,0,\pi)$	$(0,0,0)$	$(\pi,0,\pi,0)$	$(0,\pi,0)$	$(0,0,\pi,\pi)$
14	$(\epsilon^2,\epsilon,1)$	$(\epsilon^2,\epsilon^2,\frac{\pi}{4})$	$(\epsilon,0,\epsilon^2)$	$(\frac{\pi}{4},\epsilon^2,0)$	$(\epsilon^2,\frac{\pi}{4},0)$	$(\epsilon^2,\frac{\pi}{4},\epsilon)$
	$(\epsilon^2,\epsilon,1)$	(π,π,π)	$(0,0,\pi)$	$(0,0,0,0)$	$(0,0,\pi)$	$(0,0,\pi,0)$
15	$(\epsilon^2,\epsilon,1)$	$(\epsilon^2,\epsilon^2,\frac{\pi}{4})$	$(\epsilon,0,\epsilon^2)$	$(\frac{\pi}{4},0,\epsilon^2)$	$(\epsilon^2,\epsilon,\epsilon^2)$	$(\epsilon,\epsilon,\epsilon^2)$
	$(\epsilon^2,\epsilon,1)$	(π,π,π)	$(0,0,0)$	$(0,0,0,\pi)$	$(0,0,0)$	$(\pi,0,0,\pi)$
16	$(\epsilon,1,\epsilon)$	$(\epsilon,\epsilon^2,\frac{\pi}{4})$	$(\epsilon,0,\epsilon^2)$	$(\epsilon,\frac{\pi}{4},\epsilon^2)$	$(\frac{\pi}{4},0,0)$	$(\frac{\pi}{4},\epsilon^2,\epsilon^2)$
	$(\epsilon,1,1)$	$(\pi,0,0)$	$(0,0,0)$	$(\pi,0,0,\pi)$	$(0,0,\pi)$	$(0,0,0,0)$
17	$(\epsilon^2,\epsilon,1)$	$(\epsilon^2,\epsilon^2,\frac{\pi}{4})$	$(\epsilon,0,\epsilon^2)$	$(\frac{\pi}{4},0,\epsilon^2)$	$(\epsilon^2,0,\epsilon^2)$	$(\epsilon,0,\epsilon^2)$
	$(\epsilon^2,\epsilon,1)$	(π,π,π)	$(0,0,0)$	$(0,0,0,\pi)$	$(0,0,0)$	$(0,0,0,\pi)$
18	$(\epsilon^2,\epsilon,1)$	$(\epsilon^2,\epsilon^2,\epsilon^2)$	$(\epsilon,0,\epsilon^2)$	$(\frac{\pi}{4},\epsilon^2,\frac{\pi}{4})$	$(\epsilon,\epsilon^2,\epsilon^2)$	$(0,\epsilon^2,0)$
	$(\epsilon^2,\epsilon,1)$	$(\pi,0,\pi)$	$(0,0,0)$	$(\pi,0,\pi,0)$	$(\pi,0,\pi)$	$(0,0,0,0)$
19	$(\epsilon,1,\epsilon)$	$(\epsilon^2,\epsilon^2,\epsilon^2)$	$(\frac{\pi}{4},0,\epsilon^2)$	$(\epsilon^2,\frac{\pi}{4},\frac{\pi}{4})$	$(0,\epsilon,\epsilon)$	$(\epsilon^2,\epsilon^2,\frac{\pi}{4})$
	$(\epsilon,1,1)$	(π,π,π)	$(0,0,0)$	$(\pi,0,0,0)$	$(0,0,\pi)$	$(0,0,0,0)$
20	$(\epsilon^2,\epsilon,1)$	$(\epsilon^2,\epsilon^2,\epsilon^2)$	$(\frac{\pi}{4},0,0)$	$(\frac{\pi}{4},\epsilon^2,\frac{\pi}{4})$	$(\epsilon^2,\epsilon^2,\epsilon^2)$	$(\epsilon,\epsilon^2,\epsilon^2)$
	$(\epsilon^2,\epsilon,1)$	$(0,\pi,\pi)$	$(0,0,0)$	$(\pi,0,0,\pi)$	$(0,0,0)$	$(\pi,0,0,\pi)$
21	$(\epsilon,1,\epsilon)$	$(\epsilon^2,\epsilon^2,\epsilon^2)$	$(\frac{\pi}{4},0,0)$	$(0,\epsilon^2,\frac{\pi}{4})$	$(\epsilon,\frac{\pi}{4},\epsilon)$	$(\epsilon^2,\epsilon,\epsilon^2)$
	$(\epsilon,1,1)$	(π,π,π)	$(0,0,0)$	$(0,0,0,0)$	$(0,0,0)$	$(\pi,0,\pi,0)$
22	$(\epsilon^2,\epsilon,1)$	$(\epsilon^2,\epsilon^2,0)$	$(\frac{\pi}{4},0,\epsilon)$	$(\frac{\pi}{4},\epsilon^2,\frac{\pi}{4})$	$(\epsilon^2,\frac{\pi}{4},\epsilon)$	$(\epsilon^2,\frac{\pi}{4},0)$
	$(\epsilon^2,\epsilon,1)$	$(0,0,0)$	$(0,0,0)$	$(\pi,0,0,0)$	$(0,\pi,0)$	$(0,0,\pi,0)$

Table A.2: Supplementary information for the reconstruction of the mass matrices of the models in Table 6.1 ($\varphi_i^\ell = \varphi_i^{\ell'} = \varphi_i^{D'} = \alpha_j^D = 0$ for $i = 1, 2, 3$ and $j = 1, 2$).

List of Figures

1.1 Different roads to Grand Unification 3

2.1 Possible neutrino mass orderings 8

4.1 Froggatt-Nielsen mechanism 19
4.2 Stability of sum rules 25

5.1 Procedure for obtaining the seesaw realizations and texture sets 31
5.2 Distributions of mass hierarchies in M_D and M_R 41
5.3 Distributions of mixings in U_D, $U_{D'}$, U_R, and U_ℓ 41
5.4 Fraction of special cases of all seesaw realizations 42
5.5 χ^2 distribution of valid seesaw realizations 43
5.6 Yukawa coupling distribution of M_ℓ, M_D, and M_R 44
5.7 Scheme for obtaining seesaw realizations and texture sets in EQLC . . . 45

6.1 Overview of lepton flavor models for varying flavor group 54

7.1 $\text{Br}(\mu \to e\gamma)$, $\text{Br}(\tau \to e\gamma)$, and $\text{Br}(\tau \to \mu\gamma)$ for 72 CP conserving seesaw realizations 59
7.2 $\text{Br}(\mu \to e\gamma)$ as a function of m_3^R for texture #1 60
7.3 $\text{Br}(\mu \to e\gamma)$ as function of the PMNS mixing parameters for the complexified texture #1 61

8.1 Overview of $SU(5) \times G_A$ models for varying flavor group 67
8.2 5D SUSY $SU(5)$ GUT setup on two 5D throats 68
8.3 Local flavor symmetry breaking $G_A \ltimes G_B \to G_A$ 69
8.4 Effect of the non-Abelian flavor symmetry on θ_{23} 70

List of Tables

5.1	72 selected seesaw realizations	40
6.1	22 lepton flavor models	53
7.1	Current and future sensitivities of direct experimental LFV searches	56
A.1	Supplementary information for seesaw realizations in Table 5.1	80
A.2	Supplementary information for the models in Table 6.1	81

Bibliography

[1] "The Science Ahead, the Way to Discovery," 2005. The High-Energy Physics Advisory Panel long range plan for U.S.

[2] **Super-Kamiokande** Collaboration, S. Fukuda *et al.*, "Determination of solar neutrino oscillation parameters using 1496 days of super-kamiokande-i data," *Phys. Lett.* **B539** (2002) 179–187, `hep-ex/0205075`.

[3] **Super-Kamiokande** Collaboration, Y. Fukuda *et al.*, "Evidence for oscillation of atmospheric neutrinos," *Phys. Rev. Lett.* **81** (1998) 1562–1567, `hep-ex/9807003`.

[4] **KamLAND** Collaboration, T. Araki *et al.*, "Measurement of neutrino oscillation with kamland: Evidence of spectral distortion," *Phys. Rev. Lett.* **94** (2005) 081801, `hep-ex/0406035`.

[5] **K2K** Collaboration, E. Aliu *et al.*, "Evidence for muon neutrino oscillation in an accelerator- based experiment," *Phys. Rev. Lett.* **94** (2005) 081802, `hep-ex/0411038`.

[6] H. Georgi and S. L. Glashow, "Unity of All Elementary Particle Forces," *Phys. Rev. Lett.* **32** (1974) 438–441.

[7] H. Georgi, "Unified Gauge Theories," 1975. In *Coral Gables, Proceedings, Theories and Experiments In High Energy Physics*, New York 1975, 329–339.

[8] J. C. Pati and A. Salam, "Lepton Number as the Fourth Color," *Phys. Rev.* **D10** (1974) 275–289.

[9] P. Minkowski, "mu $\to$ e gamma at a Rate of One Out of 1-Billion Muon Decays?," *Phys. Lett.* **B67** (1977) 421.

[10] T. Yanagida, "Horizontal gauge symmetry and masses of neutrinos," 1979. In Proceedings of the Workshop on the Baryon Number of the Universe and Unified Theories, Tsukuba, Japan, 13-14 Feb.

[11] M. Gell-Mann, P. Ramond, and R. Slansky, "Complex spinors and unified theories," 1979. Print-80-0576 (CERN).

[12] S. L. Glashow, "Proceedings of the Cargese Summer Institute on Quarks and Leptons," 1979. New York, 1980.

[13] M. Magg and C. Wetterich, "Neutrino mass problem and gauge hierarchy," *Phys. Lett.* **B94** (1980) 61.

[14] R. N. Mohapatra and G. Senjanovic, "Neutrino mass and spontaneous parity nonconservation," *Phys. Rev. Lett.* **44** (1980) 912.

[15] R. N. Mohapatra and G. Senjanovic, "Neutrino Masses and Mixings in Gauge Models with Spontaneous Parity Violation," *Phys. Rev.* **D23** (1981) 165.

[16] J. Schechter and J. W. F. Valle, "Neutrino Masses in SU(2) x U(1) Theories," *Phys. Rev.* **D22** (1980) 2227.

[17] G. Lazarides, Q. Shafi, and C. Wetterich, "Proton Lifetime and Fermion Masses in an SO(10) Model," *Nucl. Phys.* **B181** (1981) 287.

[18] H. Georgi, H. R. Quinn, and S. Weinberg, "Hierarchy of Interactions in Unified Gauge Theories," *Phys. Rev. Lett.* **33** (1974) 451–454.

[19] S. Dimopoulos, S. Raby, and F. Wilczek, "Supersymmetry and the Scale of Unification," *Phys. Rev.* **D24** (1981) 1681–1683.

[20] S. Dimopoulos and H. Georgi, "Softly Broken Supersymmetry and SU(5)," *Nucl. Phys.* **B193** (1981) 150.

[21] B. Pontecorvo, "Mesonium and antimesonium," *Sov. Phys. JETP* **6** (1957) 429.

[22] Z. Maki, M. Nakagawa, and S. Sakata, "Remarks on the unified model of elementary particles," *Prog. Theor. Phys.* **28** (1962) 870.

BIBLIOGRAPHY

[23] P. F. Harrison, D. H. Perkins, and W. G. Scott, "A redetermination of the neutrino mass-squared difference in tri-maximal mixing with terrestrial matter effects," *Phys. Lett.* **B458** (1999) 79–92, arXiv:hep-ph/9904297.

[24] F. Plentinger and W. Rodejohann, "Deviations from TribiMaximal Neutrino Mixing," *Phys. Lett.* **B625** (2005) 264–276, hep-ph/0507143.

[25] D. Majumdar and A. Ghosal, "Probing deviations from tri-bimaximal mixing through ultra high energy neutrino signals," *Phys. Rev.* **D75** (2007) 113004, arXiv:hep-ph/0608334.

[26] Z.-z. Xing, "Nearly tri-bimaximal neutrino mixing and CP violation," *Phys. Lett.* **B533** (2002) 85–93, arXiv:hep-ph/0204049.

[27] N. Cabibbo, "Unitary symmetry and leptonic decays," *Phys. Rev. Lett.* **10** (1963) 531–532.

[28] M. Kobayashi and T. Maskawa, "CP Violation in the Renormalizable Theory of Weak Interaction," *Prog. Theor. Phys.* **49** (1973) 652–657.

[29] E. Ma, "Polygonal Derivation of the Neutrino Mass Matrix," hep-ph/0409288.

[30] C. Hagedorn, M. Lindner, and F. Plentinger, "The discrete flavor symmetry D(5)," *Phys. Rev.* **D74** (2006) 025007, arXiv:hep-ph/0604265.

[31] E. Ma and G. Rajasekaran, "Softly Broken A(4) Symmetry for Nearly Degenerate Neutrino Masses," *Phys. Rev.* **D64** (2001) 113012, hep-ph/0106291.

[32] K. S. Babu, E. Ma, and J. W. F. Valle, "Underlying A(4) Symmetry for the Neutrino Mass Matrix and the Quark Mixing Matrix," *Phys. Lett.* **B552** (2003) 207–213, hep-ph/0206292.

[33] M. Hirsch, J. C. Romao, S. Skadhauge, J. W. F. Valle, and A. Villanova del Moral, "Phenomenological Tests of Supersymmetric A(4) Family Symmetry Model of Neutrino Mass," *Phys. Rev.* **D69** (2004) 093006, hep-ph/0312265.

[34] P. H. Frampton and T. W. Kephart, "Simple nonAbelian finite flavor groups and fermion masses," *Int. J. Mod. Phys.* **A10** (1995) 4689–4704, arXiv:hep-ph/9409330.

[35] A. Aranda, C. D. Carone, and R. F. Lebed, "Maximal Neutrino Mixing from a Minimal Flavor Symmetry," *Phys. Rev.* **D62** (2000) 016009, hep-ph/0002044.

[36] P. D. Carr and P. H. Frampton, "Group theoretic bases for tribimaximal mixing," arXiv:hep-ph/0701034.

[37] A. Aranda, "Neutrino mixing from the double tetrahedral group T'," *Phys. Rev.* **D76** (2007) 111301, arXiv:0707.3661 [hep-ph].

[38] G. Altarelli, "Models of neutrino masses and mixings: A progress report," arXiv:0705.0860 [hep-ph].

[39] A. Y. Smirnov, "Neutrinos: '...annus mirabilis'," hep-ph/0402264.

[40] M. Raidal, "Prediction theta(c) + theta(sol) = pi/4 from flavor physics: A new evidence for grand unification?," *Phys. Rev. Lett.* **93** (2004) 161801, hep-ph/0404046.

[41] H. Minakata and A. Y. Smirnov, "Neutrino Mixing and Quark Lepton Complementarity," *Phys. Rev.* **D70** (2004) 073009, hep-ph/0405088.

[42] H. Fritzsch and Z.-z. Xing, "Mass and flavor mixing schemes of quarks and leptons," *Prog. Part. Nucl. Phys.* **45** (2000) 1–81, arXiv:hep-ph/9912358.

[43] P. H. Frampton, S. L. Glashow, and D. Marfatia, "Zeroes of the Neutrino Mass Matrix," *Phys. Lett.* **B536** (2002) 79–82, hep-ph/0201008.

[44] A. Ibarra and G. G. Ross, "Neutrino Properties from Yukawa Structure," *Phys. Lett.* **B575** (2003) 279–289, hep-ph/0307051.

[45] W. Grimus, A. S. Joshipura, L. Lavoura, and M. Tanimoto, "Symmetry Realization of Texture Zeros," *Eur. Phys. J.* **C36** (2004) 227–232, hep-ph/0405016.

[46] S. Kaneko, H. Sawanaka, and M. Tanimoto, "Hybrid Textures of Neutrinos," *JHEP* **08** (2005) 073, hep-ph/0504074.

[47] C. Hagedorn and W. Rodejohann, "Minimal Mass Matrices for Dirac Neutrinos," *JHEP* **07** (2005) 034, hep-ph/0503143.

[48] C. D. Froggatt and H. B. Nielsen, "Hierarchy of quark masses, cabibbo angles and cp violation," *Nucl. Phys.* **B147** (1979) 277.

BIBLIOGRAPHY

[49] H. Fusaoka and Y. Koide, "Updated estimate of running quark masses," *Phys. Rev.* **D57** (1998) 3986–4001, arXiv:hep-ph/9712201.

[50] V. Barger, D. Marfatia, and K. Whisnant, "Progress in the Physics of Massive Neutrinos," *Int. J. Mod. Phys.* **E12** (2003) 569–647, hep-ph/0308123.

[51] T. Schwetz, "Global fits to neutrino oscillation data," *Phys. Scripta* **T127** (2006) 1–5, hep-ph/0606060.

[52] M. Jamin, "Quark masses," *Granada* (2006).

[53] E. Blucher *et al.*, "Status of the Cabibbo angle," arXiv:hep-ph/0512039.

[54] R. Gatto, G. Sartori, and M. Tonin, "Weak Selfmasses, Cabibbo Angle, and Broken SU(2) x SU(2)," *Phys. Lett.* **B28** (1968) 128–130.

[55] R. J. Oakes, "SU(2) x SU(2) breaking and the Cabibbo angle," *Phys. Lett.* **B29** (1969) 683–685.

[56] H. Fritzsch, "Quark Masses and Flavor Mixing," *Nucl. Phys.* **B155** (1979) 189.

[57] L. J. Hall and A. Rasin, "On the generality of certain predictions for quark mixing," *Phys. Lett.* **B315** (1993) 164–169, arXiv:hep-ph/9303303.

[58] R. Barbieri, L. J. Hall, and A. Romanino, "Precise tests of a quark mass texture," *Nucl. Phys.* **B551** (1999) 93–101, arXiv:hep-ph/9812384.

[59] R. G. Roberts, A. Romanino, G. G. Ross, and L. Velasco-Sevilla, "Precision test of a Fermion mass texture," *Nucl. Phys.* **B615** (2001) 358–384, arXiv:hep-ph/0104088.

[60] H. Georgi and C. Jarlskog, "A New Lepton - Quark Mass Relation in a Unified Theory," *Phys. Lett.* **B86** (1979) 297–300.

[61] S. Pakvasa and H. Sugawara, "Discrete Symmetry and Cabibbo Angle," *Phys. Lett.* **B73** (1978) 61.

[62] S. Pakvasa and H. Sugawara, "Mass of the t Quark in SU(2) x U(1)," *Phys. Lett.* **B82** (1979) 105.

[63] Y. Yamanaka, H. Sugawara, and S. Pakvasa, "Permutation Symmetries and the Fermion Mass Matrix," *Phys. Rev.* **D25** (1982) 1895.

[64] T. Brown, N. Deshpande, S. Pakvasa, and H. Sugawara, "CP Nonconservation and Rare Processes in S(4) Model of Permutation Symmetry," *Phys. Lett.* **B141** (1984) 95.

[65] H. Harari, H. Haut, and J. Weyers, "Quark Masses and Cabibbo Angles," *Phys. Lett.* **B78** (1978) 459.

[66] Y. Koide, "Fermion - Boson Two-Body Model of Quarks and Leptons and Cabibbo Mixing," *Nuovo Cim. Lett.* **34** (1982) 201.

[67] Y. Koide, "A New View of Quark and Lepton Mass Hierarchy," *Phys. Rev.* **D28** (1983) 252.

[68] Y. Koide, "A Fermion - Boson Composite Model of Quarks and Leptons," *Phys. Lett.* **B120** (1983) 161.

[69] T. Yamashita, "Anomalous U(1) GUT," `hep-ph/0503265`.

[70] F. Plentinger, "Discrete flavour symmetries," dec, 2005.

[71] F. Plentinger, G. Seidl, and W. Winter, "Systematic parameter space search of extended quark-lepton complementarity," *Nucl. Phys.* **B791** (2008) 60–92, `arXiv:hep-ph/0612169`.

[72] S. Antusch, P. Huber, J. Kersten, T. Schwetz, and W. Winter, "Is there maximal mixing in the lepton sector?," *Phys. Rev.* **D70** (2004) 097302, `hep-ph/0404268`.

[73] R. Gandhi and W. Winter, "Physics with a very long neutrino factory baseline," *Phys. Rev.* **D75** (2007) 053002, `arXiv:hep-ph/0612158`.

[74] S. Niehage and W. Winter, "Entangled maximal mixings in $U_{\text{PMNS}} = U_l^\dagger U_\nu$, and a connection to complex mass textures," *Phys. Rev.* **D78** (2008) 013007, `arXiv:0804.1546 [hep-ph]`.

[75] F. Plentinger, G. Seidl, and W. Winter, "The Seesaw Mechanism in Quark-Lepton Complementarity," *Phys. Rev.* **D76** (2007) 113003, `arXiv:0707.2379 [hep-ph]`.

BIBLIOGRAPHY

[76] F. Plentinger, G. Seidl, and W. Winter.
http://theorie.physik.uni-wuerzburg.de/~winter/Resources/Textures/index.html.

[77] F. Plentinger, "Constructing Textures in Extended Quark-Lepton Complementarity," arXiv:0709.1949 [hep-ph].

[78] M. Jezabek and Y. Sumino, "Neutrino masses and bimaximal mixing," *Phys. Lett.* **B457** (1999) 139–146, hep-ph/9904382.

[79] C. Giunti and M. Tanimoto, "CP violation in bilarge lepton mixing," *Phys. Rev.* **D66** (2002) 113006, arXiv:hep-ph/0209169.

[80] P. H. Frampton, S. T. Petcov, and W. Rodejohann, "On deviations from bimaximal neutrino mixing," *Nucl. Phys.* **B687** (2004) 31–54, hep-ph/0401206.

[81] T. Ohlsson, "Bimaximal Fermion Mixing from the Quark and Leptonic Mixing Matrices," *Phys. Lett.* **B622** (2005) 159–164, hep-ph/0506094.

[82] S. Antusch and S. F. King, "Charged lepton corrections to neutrino mixing angles and CP phases revisited," *Phys. Lett.* **B631** (2005) 42–47, arXiv:hep-ph/0508044.

[83] K. Cheung, S. K. Kang, C. S. Kim, and J. Lee, "Lepton Flavor Violation As a Probe of Quark-Lepton Unification," *Phys. Rev.* **D72** (2005) 036003, hep-ph/0503122.

[84] K. A. Hochmuth and W. Rodejohann, "Low and High Energy Phenomenology of Quark-Lepton Complementarity Scenarios," *Phys. Rev.* **D75** (2007) 073001, arXiv:hep-ph/0607103.

[85] W. Rodejohann, "A parametrization for the neutrino mixing matrix," *Phys. Rev.* **D69** (2004) 033005, hep-ph/0309249.

[86] N. Li and B.-Q. Ma, "Unified Parametrization of Quark and Lepton Mixing Matrices," *Phys. Rev.* **D71** (2005) 097301, hep-ph/0501226.

[87] Z.-z. Xing, "Nontrivial Correlation Between the CKM and MNS Matrices," *Phys. Lett.* **B618** (2005) 141–149, hep-ph/0503200.

[88] A. Datta, L. Everett, and P. Ramond, "Cabibbo Haze in Lepton Mixing," *Phys. Lett.* **B620** (2005) 42–51, hep-ph/0503222.

[89] L. L. Everett, "Viewing lepton mixing through the cabibbo haze," *Phys. Rev.* **D73** (2006) 013011, hep-ph/0510256.

[90] B. C. Chauhan, M. Picariello, J. Pulido, and E. Torrente-Lujan, "Quark-lepton complementarity, neutrino and standard model data predict ($\theta_{13}^{PMNS} = 9^{+1}_{-2})°$," *Eur. Phys. J.* **C50** (2007) 573–578, arXiv:hep-ph/0605032.

[91] A. Dighe, S. Goswami, and P. Roy, "Quark-lepton complementarity with quasidegenerate Majorana neutrinos," *Phys. Rev.* **D73** (2006) 071301, arXiv:hep-ph/0602062.

[92] M. A. Schmidt and A. Y. Smirnov, "Quark lepton complementarity and renormalization group effects," *Phys. Rev.* **D74** (2006) 113003, arXiv:hep-ph/0607232.

[93] T. Ohlsson and G. Seidl, "A flavor symmetry model for bilarge leptonic mixing and the lepton masses," *Nucl. Phys.* **B643** (2002) 247–279, hep-ph/0206087.

[94] P. H. Frampton and R. N. Mohapatra, "Possible gauge theoretic origin for quark-lepton complementarity," *JHEP* **01** (2005) 025, hep-ph/0407139.

[95] S. Antusch, S. F. King, and R. N. Mohapatra, "Quark lepton complementarity in unified theories," *Phys. Lett.* **B618** (2005) 150–161, hep-ph/0504007.

[96] M. Picariello, "Neutrino CP violating parameters from nontrivial quark- lepton correlation: A S(3) x GUT model," *Int. J. Mod. Phys.* **A23** (2008) 4435–4448, arXiv:hep-ph/0611189.

[97] A. Hernandez-Galeana, "Fermion masses and Neutrino mixing in an $U(1)_H$ flavor symmetry model with hierarchical radiative generation for light charged fermion masses," *Phys. Rev.* **D76** (2007) 093006, arXiv:0710.2834 [hep-ph].

[98] P. Huber, M. Lindner, M. Rolinec, T. Schwetz, and W. Winter, "Prospects of accelerator and reactor neutrino oscillation experiments for the coming ten years," *Phys. Rev.* **D70** (2004) 073014, hep-ph/0403068.

[99] H. Minakata, H. Nunokawa, W. J. C. Teves, and R. Zukanovich Funchal, "Reactor measurement of theta(12): Principles, accuracies and physics potentials," *Phys. Rev.* **D71** (2005) 013005, hep-ph/0407326.

BIBLIOGRAPHY

[100] A. Bandyopadhyay, S. Choubey, S. Goswami, and S. T. Petcov, "High precision measurements of theta(solar) in solar and reactor neutrino experiments," *Phys. Rev.* **D72** (2005) 033013, hep-ph/0410283.

[101] V. Barger, P. Huber, D. Marfatia, and W. Winter, "Upgraded experiments with super neutrino beams: Reach versus Exposure," *Phys. Rev.* **D76** (2007) 031301, arXiv:hep-ph/0610301.

[102] A. Cervera *et al.*, "Golden measurements at a neutrino factory," *Nucl. Phys.* **B579** (2000) 17–55, hep-ph/0002108.

[103] P. Huber, M. Lindner, and W. Winter, "Superbeams versus neutrino factories," *Nucl. Phys.* **B645** (2002) 3–48, hep-ph/0204352.

[104] P. Huber, M. Lindner, M. Rolinec, and W. Winter, "Optimization of a neutrino factory oscillation experiment," *Phys. Rev.* **D74** (2006) 073003, arXiv:hep-ph/0606119.

[105] W. Winter, "Neutrino Oscillation Observables from Mass Matrix Structure," *Phys. Lett.* **B659** (2008) 275–280, arXiv:0709.2163 [hep-ph].

[106] S. Pascoli, S. T. Petcov, and A. Riotto, "Connecting low energy leptonic CP-violation to leptogenesis," *Phys. Rev.* **D75** (2007) 083511, arXiv:hep-ph/0609125.

[107] H. Arason *et al.*, "Top quark and Higgs mass bounds from a numerical study of superGUTs," *Phys. Rev. Lett.* **67** (1991) 2933.

[108] H. Arason, D. J. Castano, E. J. Piard, and P. Ramond, "Mass and mixing angle patterns in the standard model and its minimal supersymmetric extension," *Phys. Rev.* **D47** (1993) 232–240, arXiv:hep-ph/9204225.

[109] A. Dighe, S. Goswami, and P. Roy, "Radiatively broken symmetries of nonhierarchical neutrinos," *Phys. Rev.* **D76** (2007) 096005, arXiv:0704.3735 [hep-ph].

[110] F. Plentinger, G. Seidl, and W. Winter, "Group Space Scan of Flavor Symmetries for Nearly Tribimaximal Lepton Mixing," *JHEP* **04** (2008) 077, arXiv:0802.1718 [hep-ph].

[111] L. M. Krauss and F. Wilczek, "Discrete Gauge Symmetry in Continuum Theories," *Phys. Rev. Lett.* **62** (1989) 1221.

[112] F. Plentinger and G. Seidl, "Mapping out SU(5) GUTs with non-Abelian discrete flavor symmetries," *Phys. Rev.* **D78** (2008) 045004, arXiv:0803.2889 [hep-ph].

[113] F. Deppisch, F. Plentinger, W. Porod, R. Ruckl, and G. Seidl, "Lepton Flavor and CP Violation in mSUGRA (in preparation),".

[114] **Particle Data Group** Collaboration, S. Eidelman *et al.*, "Review of particle physics," *Phys. Lett.* **B592** (2004) 1.

[115] **BABAR** Collaboration, B. Aubert *et al.*, "Search for lepton flavor violation in the decay $\tau \to \mu\gamma$," *Phys. Rev. Lett.* **95** (2005) 041802, arXiv:hep-ex/0502032.

[116] J. Hisano and D. Nomura, "Solar and atmospheric neutrino oscillations and lepton flavor violation in supersymmetric models with the right-handed neutrinos," *Phys. Rev.* **D59** (1999) 116005, arXiv:hep-ph/9810479.

[117] F. Deppisch, H. Pas, A. Redelbach, R. Ruckl, and Y. Shimizu, "Probing the Majorana mass scale of right-handed neutrinos in mSUGRA," *Eur. Phys. J.* **C28** (2003) 365–374, arXiv:hep-ph/0206122.

[118] J. A. Casas and A. Ibarra, "Oscillating neutrinos and mu -> e, gamma," *Nucl. Phys.* **B618** (2001) 171–204, arXiv:hep-ph/0103065.

[119] G. Cacciapaglia, C. Csaki, C. Grojean, and J. Terning, "Field theory on multi-throat backgrounds," *Phys. Rev.* **D74** (2006) 045019, arXiv:hep-ph/0604218.

[120] K. Agashe, A. Falkowski, I. Low, and G. Servant, "KK Parity in Warped Extra Dimension," *JHEP* **04** (2008) 027, arXiv:0712.2455 [hep-ph].

[121] D. E. Kaplan and T. M. P. Tait, "New tools for fermion masses from extra dimensions," *JHEP* **11** (2001) 051, arXiv:hep-ph/0110126.

[122] Y. Kawamura, "Triplet-doublet splitting, proton stability and extra dimension," *Prog. Theor. Phys.* **105** (2001) 999–1006, arXiv:hep-ph/0012125.

[123] Y. Nomura, "Strongly coupled grand unification in higher dimensions," *Phys. Rev.* **D65** (2002) 085036, arXiv:hep-ph/0108170.

[124] G. Seidl, "Unified model of fermion masses with Wilson line flavor symmetry breaking," arXiv:0811.3775 [hep-ph].

[125] K. S. Babu and S. M. Barr, "Natural suppression of Higgsino mediated proton decay in supersymmetric SO(10)," *Phys. Rev.* **D48** (1993) 5354–5364, arXiv:hep-ph/9306242.

[126] K. Kurosawa, N. Maru, and T. Yanagida, "Nonanomalous R-symmetry in supersymmetric unified theories of quarks and leptons," *Phys. Lett.* **B512** (2001) 203–210, arXiv:hep-ph/0105136.

I want morebooks!

Buy your books fast and straightforward online - at one of world's fastest growing online book stores! Environmentally sound due to Print-on-Demand technologies.

Buy your books online at
www.morebooks.shop

Kaufen Sie Ihre Bücher schnell und unkompliziert online – auf einer der am schnellsten wachsenden Buchhandelsplattformen weltweit! Dank Print-On-Demand umwelt- und ressourcenschonend produziert.

Bücher schneller online kaufen
www.morebooks.shop

KS OmniScriptum Publishing
Brivibas gatve 197
LV-1039 Riga, Latvia
Telefax: +371 686 204 55

info@omniscriptum.com
www.omniscriptum.com

Printed by Books on Demand GmbH, Norderstedt / Germany